Introduction to FUNGI

Jagdish Acharya

RAJAT PUBLICATIONS
New Delhi

Introduction to Fungi
by *Jagdish Acharya*

This edition is published by Rajat Publications
4740/23, Ansari Road Daryaganj,
New Delhi - 110002

ISBN: 978-81-7880-369-2

Edition 2025

Printed and bound in India

RAJAT
PUBLICATIONS

4740/23, Ansari Road, Daryaganj,
New Delhi- 110002 (India)
Mob:- +91-9811484973, +91-9811105389
E-mail: rajat_publication@yahoo.com

Preface

Fungi constitute an important group of cryptogams whose study is essential for all the students of Botany. Fungi constitute an integral part of the biological spectrum having far reaching influence on other components of environment. From the point of human beings fungi are both friends as well as foes. In the evolutionary hierarchy of plants, fungi occupy a pivotal position as they pave the path for structural and functional adaptations of plants at the basic level without the help of photosynthetic mechanism. A detailed study of fungi is a part and parcel of the curriculum for the students of Botany at the undergraduate and post graduate level.

Jagdish Acharya

Contents

1

Introduction to the Fungi

The Kingdom Fungi includes some of the most important organisms, both in terms of their ecological and economic roles. By breaking down dead organic material, they continue the cycle of nutrients through ecosystems. In addition, most vascular plants could not grow without the symbiotic fungi, or mycorrhizae, that inhabit their roots and supply essential nutrients. Other fungi provide numerous drugs (such as penicillin and other antibiotics), foods like mushrooms, truffles and morels, and the bubbles in bread, champagne, and beer.

Fungi also cause a number of plant and animal diseases: in humans, ringworm, athlete's foot, and several more serious diseases are caused by fungi. Because fungi are more chemically and genetically similar to animals than other organisms, this makes fungal diseases very difficult to treat. Plant diseases caused by fungi include rusts, smuts, and leaf, root, and stem rots, and may cause severe damage to crops. However, a number of fungi, in particular the yeasts, are important "model organisms" for studying problems in genetics and molecular biology.

The organisms of the fungal lineage include mushrooms, rusts, smuts, puffballs, truffles, morels, molds, and yeasts, as well as many less well-known organisms. More than 70,000 species of fungi have been described; however, some estimates of total numbers suggest that 1.5 million species may exist.

As the sister group of animals and part of the eukaryotic crown group that radiated about a billion years ago, the fungi constitute an independent group equal in rank to that of plants and animals. They share with animals the ability to export hydrolytic enzymes that break down biopolymers, which can be absorbed for nutrition. Rather than requiring a stomach to accomplish digestion, fungi live in their own food supply and simply grow into new food as the local environment becomes nutrient depleted.

Most biologists have seen dense filamentous fungal colonies growing on rich nutrient agar plates, but in nature the filaments can be much longer and the colonies less dense. When one of the filaments contacts a food supply, the entire colony mobilizes and reallocates resources to exploit the new food. Should all food become depleted, sporulation is triggered. Although the fungal filaments and spores are microscopic, the colony can be very large with individuals of some species rivaling the mass of the largest animals or plants.

Prior to mating in sexual reproduction, individual fungi communicate with other individuals chemically via pheromones. In every phylum at least one pheromone has been characterized, and they range from sesquiterpines and derivatives of the carotenoid pathway in chytridiomycetes and zygomycetes to oligopeptides in ascomycetes and basidiomycetes.

Within their varied natural habitats fungi usually are the primary decomposer organisms present. Many species are free-living saprobes (users of carbon fixed by other organisms) in woody substrates, soils, leaf litter, dead animals, and animal exudates. The large cavities eaten out of living trees by wood-decaying fungi provide nest holes for a variety of animals, and extinction of the ivory billed woodpecker was due in large part to loss, through human activity, of nesting trees in bottom land hardwoods. In some low nitrogen environments several independent groups of fungi have adaptations such as nooses and sticky knobs with which to trap and degrade nematodes and other small animals. A number of references on fungal ecology are available.

However, many other fungi are biotrophs, and in this role a number of successful groups form symbiotic associations with plants (including algae), animals (especially arthropods), and prokaryotes. Although lichens may seem infrequent in polluted cities, they can form the dominant vegetation in nordic environments, and there is a better than 80 per cent chance that any plant you find is mycorrhizal. Leaf and stem endophytes are a more recent discovery, and some of these fungi can protect the plants they inhabit from herbivory and even influence flowering and other aspects of plant reproductive biology. Fungi are our most important plant pathogens, and include rusts, smuts, and many ascomycetes such as the agents of Dutch elm disease and chestnut blight. Among the other well known associations are fungal parasites of animals. Humans, may succumb to diseases caused by *Pneumocystis* (a type of pneumonia that affects individuals with supressed immune systems), *Coccidioides* (valley fever), *Ajellomyces* (blastomycosis and histoplasmosis), and *Cryptococcus* (cryptococcosis).

Fungal spores may be actively or passively released for dispersal by several effective methods. The air we breathe is filled with spores of species that are air dispersed. These usually are species that produce large numbers of spores include many species pathogenic on agricultural crops and trees. Other species are adapted for dispersal within or on the surfaces of animals (particularly arthropods). Some fungi are rain splash or flowing water dispersed. In a few cases the forcible release of spores is sufficient to serve as the dispersal method as well. The function of some spores is not primarily for dispersal, but to allow the organisms to survive as resistant cells during periods when the conditions of the environment are not conducive to growth. A number of fungi are used in the processing and flavouring of foods and in production of antibiotics and organic acids. Other fungi produce secondary metabolites such as aflatoxins that may be potent toxins and carcinogens in food of birds, fish, humans, and other mammals.

A few species are studied as model organisms that can be used to gain knowledge of basic processes such as genetics, physiology, biochemistry, and molecular biology with results that are applicable to many organisms. Some of the fungi that have been intensively studied in this way include *Saccharomyces cerevisiae*, *Neurospora crassa*, and *Ustilago maydis*.

Most phyla appear to be terrestrial in origin, although all major groups have invaded marine and freshwater habitats An exception to this generality is the flagellum-bearing phyla Chytridiomycota, Blastocladiomycota, and Neocallimastigomycota (collectively referred to as chytrids), which probably had an aquatic origin. Extant chytrid species also occur in terrestrial environments as plant pathogenic fungi, soil fungi, and even as anaerobic inhabitants of the guts of herbivores such as cows (all Neocallimastigomycota).

FUNGI

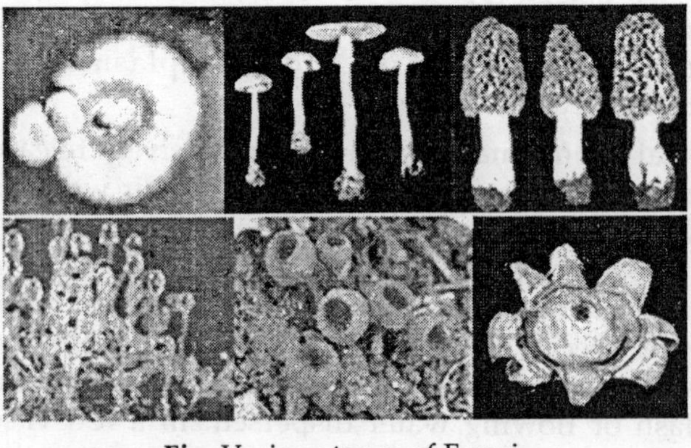

Fig. Various types of Fungi.

Fungi are living organisms that are distantly related to plants, and more closely related to animals, but rather different from either of those groups.

Fungi can be recognized by the following five characteristics:

1. The cells of fungi contain nuclei with chromosomes (like plants and animals, but unlike bacteria).
2. Fungi cannot photosynthesize (they are heterotrophic, like animals).
3. Fungi absorb their food (they are osmotrophic).
4. They mostly develop very diffuse bodies made up of a spreading network of very narrow, tubular, branching filaments called hyphae. These filaments exude enzymes, and absorb food, at their growing tips. Although these filaments are very narrow, they are collectively very long, and can explore and exploit food substrates very efficiently.
5. They usually reproduce by means of spores, which develop on, and are released by, a range of unique structures (such as mushrooms, cup fungi, and many other kinds of microscopically small fruiting bodies).

Organisms which have all five of the features just described can be found in two of the seven living Kingdoms - they make up the entire Kingdom Eumycota or true fungi, whose cell walls (the walls of the hyphae) are made largely of chitin (like the exoskeletons of insects); and part of the Kingdom Chromista, whose cell walls are made of cellulose (like plants), and which also include the brown algae - wracks and kelps (all these Chromista, though they may look very different, have similar swimming cells with two flagella at one stage in their lives).

Types of Fungi

Basidiomycetes and Ascomycetes

Fungi with sporophores (fruiting bodies) large enough to be readily visible will usually belong to one of two main groups. The Basidiomycetes or the Ascomycetes. The main difference between these two groups is in the way in which they produce their microscopic spores.

In the Basidiomycetes, the spores are produced externally, on the end of specialised cells called basidia.

In Ascomycetes, spores are produced internally, inside a sac called an ascus.

Asci and basidia are both microscopic structures.

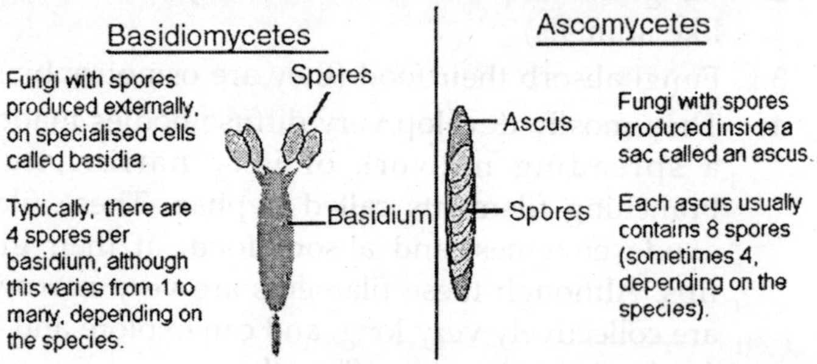

Fungi propagate by two main ways:

1. *Vegetative propagation*: fungal networks of threads (mycelium) spread outwards in soil into new territory, perhaps forming partnerships with new plants, or colonising new patches of organic matter.
2. *Spores*: Mushroom types of fungi produce massive amounts of spores on their gills. The spores are forcibly shot out into the air and are spread by the wind. Truffle fungi have below-ground fruit bodies and rely upon animals such as woylies to sniff them, dig them up, and eat them. The truffle spores pass unharmed through the animal's gut, and are deposited in the dung.

FUNGI INTERACTIONS

Fungi have a special role to play in ecosystems. Because they get their food from dead plants and animals, they are the reason that dead things decay. Fungi, therefore, are referred to as decomposers. This allows the food and nutrients to be returned to the soil and be made available to other organisms in the ecosystem. Without fungi we would be covered in piles of dead plants and animals that have never decomposed.

Fig. Relation of fungi with other organism

It is noticed that many types of toadstools appear on the ground beneath trees. The mycelium of these fungi grows in and around the growing roots of the trees. The fungus is able to get food from the tree and also allows the tree to obtain more nutrients, such as phosphorus, from the soil. This is a beneficial relationship for both the tree and the fungus. The fungus has a source of food while the tree can withstand harsher conditions with a better nutrient supply. These type of fungi are called mycorrhizae.

Some fungi can be parasites or cause diseases. They may infect plants or animals and sometimes cause their deaths. People can also get fungal infections. Tinea and ringworm are both infections of types of fungus.

When we see a toadstool or mushroom we may think that it is the main part of the fungus. This is really just the reproductive part; the main part of the fungus is under ground This main part is called the mycelium and is made up of very fine tubes called hyphae. The hyphae are only one cell wide so that is easy for each cell to obtain food.

The mushroom or toadstool part of the fungus is the fruiting body and is used for reproduction. Spores are produced in the gills of the mushroom. Air currents disperse these tiny spores. When they land in a suitable place, they germinate and grow into new hyphae.

Differences between a Mushroom and a Toadstool

The terms 'mushroom' and 'toadstool' are subjective rather than strictly scientific, and therefore are open to a range of loose interpretations. In general, fungi with fruit bodies that have a cap more or less centrally placed on top of a stem are referred to as 'mushrooms', or as 'mushrooms and toadstools'. Some people broadly consider that all fungi with a cap and stem are 'mushrooms', while others consider only edible fungi as 'mushrooms'. In the strictest sense, the word 'mushroom' refers only to members of the genus *Agaricus*, e.g. the cultivated white button mushroom. 'Toadstool' is typically applied to any fungus with a cap and stem that appears different from *Agaricus*, regardless of its edibility, or more usually if it is suspected or known to be poisonous. In a broader sense, 'toadstools' also includes other non-mushroom forms of fungi such as puffballs, earthstars, and coral fungi.

REPRODUCTION OF FUNGI

Asexual Reproduction

The ability to reproduce asexually is common to almost all fungi.

Bud Formation in Yeasts

In its simplest form asexual reproduction is by budding or binary fission. The onset of the cellular events is accompanied by the nuclear events of mitosis.

The initial events of budding can be seen as the development of a ring of chitin around the point where the bud is about to appear. This reinforces and stabilizes the cell wall. Enzymatic activity and turgor pressure the act to weaken and extrude the cell wall. New cell wall material is incorporated during this phase. Cell contents are forced into the progeny cell, and as the final phase of mitosis ends a cell plate, the point at which a new cell wall will grow inwards from, forms.

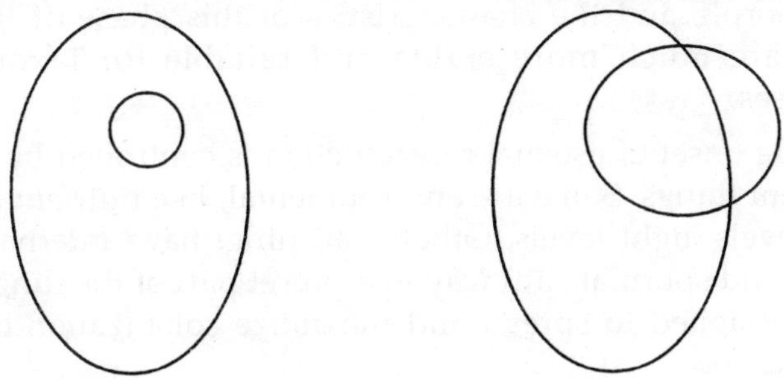

Fig. Budding of Yeast

Separation of the bud from the parent leaves a scar. When chains of yeast cells do not fully separated this can create a pseudo-mycelium.

Fragmentation

Many fungi can reproduce by fragmentation. Any mycelium that is fragmented or disrupted, provided that the fragment contains the equivalent of the peripheral growth zone, can grow into a new colony. Many fungi are sub-cultured using this hyphal fragment technique. All of this weeks practical plates have been inoculated in this way with a cork bore taken from a colonized donor plate. Cut mycelial tips do not regenerate, but branches can form some distance from the damage point.

Sporulation

By far the most important type of asexual reproduction is that of spore formation. Asexual reproduction is extremely important to fungi. It is responsible for the production of large numbers of spores throughout the year. These asexual spores are formed on a phase of the fungal life cycle termed in some texts as the mitosporic, or anamorphic phase. There can be more than one mitosporic state for each species of fungus, and in some cases the mitosporic state of very different species can look very similar. This has contributed to the problems of creating a taxonomy for the fungi that only possess mitosporic states. The sexual stage of the fungus can be termed the teleomorph, and the characteristics of this phase of the life çycle are much more stable and reliable for taxonomic purposes.

The onset of asexual reproduction is controlled by many different things. Some are environmental, like nutrient levels, CO_2 levels, light levels. Others can fungi have internal time clocks and sporulate anyway in a preset part of the fungal life cycle designed to spread and maximize colonization during one season.

Chytridiomycete Asexual Reproduction

Chytrids are quite distinct from other fungi as they have extremely simple thalli and motile zoospores. Species within this group are very simple in structure and may only consist of a single cell, perhaps with rhizoids to anchor it on to a substrate.

Asexual reproduction in the chytrids is by the production of motile zoospores, with a single, posterior flagellum, in sporangia. There will be a film of chytrid reproduction available in the practical.

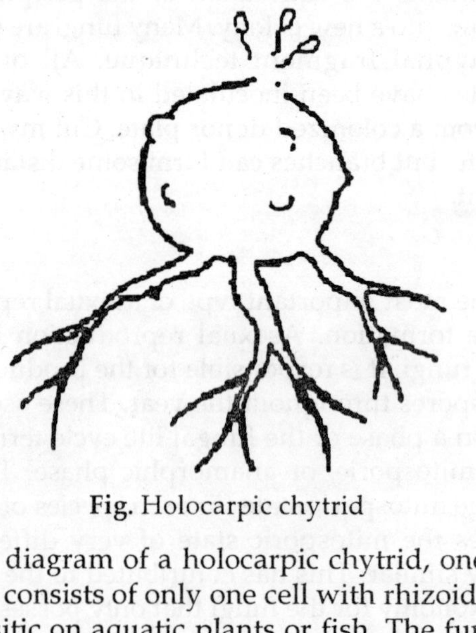

Fig. Holocarpic chytrid

This is a diagram of a holocarpic chytrid, one where the entire thallus consists of only one cell with rhizoids. These are usually parasitic on aquatic plants or fish. The fungus 'feeds' from its substrate via its rhizoids. The entire cell contents will convert to motile zoospores.

Zygomycete Asexual Reproduction

Zygomycete fungal mycelium is coenocytic. At the onset of sporulation large amounts of aerial hyphae are produced. The tips of these aerial hyphae fill with cytoplasmic contents, and the nuclei undergo repeated mitosis. Around each of the nuclei cytoplasm and organelles collect, and by the formation of copious vesicles from the Golgi, each nucleus becomes isolated from the next by a plasma membrane. Within the spaces created by this cytoplasmic cleavage, spore walls begin to form, again by the fusion of Golgi vesicles containing cell wall monomers and enzymes with the spore membrane. A sporangium forms. As these events occur so there is considerable water uptake by the forming sporangium, and as the columella forms the structure comes under considerable turgor pressure. The large sporangia can contain up to 100,000 spores.

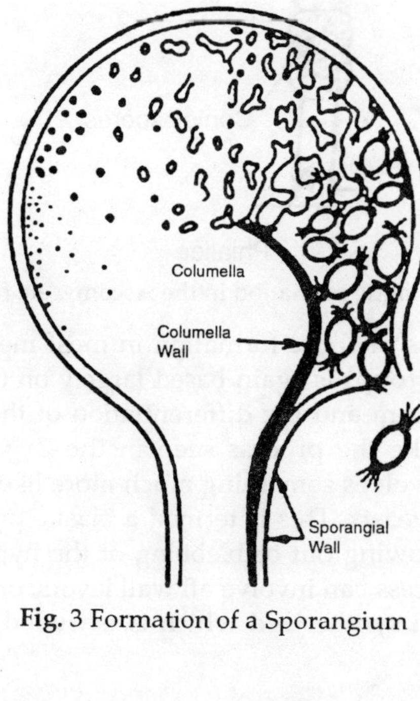

Fig. 3 Formation of a Sporangium

This diagram is redrawn from Brackers original and shows the development of a sporangium through time. As nuclei undergo repeated mitosis so the Golgi produces membrane bound vesicles filled with spore wall building materials. These coalesce around the nuclei to form a spore. Eventually they are released.

Not all sporangia are as large as this, there are many species with smaller, specialized sporangia, called sporangiola, merosporangia and some are almost conidial, forming single spored sporangia that are only distinguished from the conidium of the higher fungi by the possession of a double spore wall.

Asexual Reproduction in the Ascomycetes

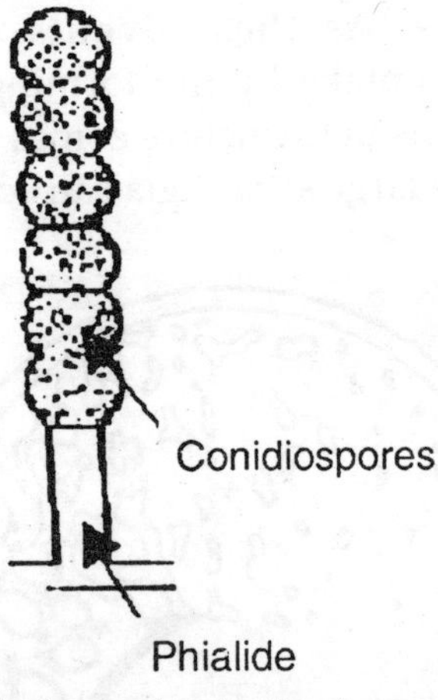

Fig. Phialide formation in the Ascomycete fungi

The process of spore formation in most members of the higher fungal groups is again based largely on the formation of aerial mycelium and the differentiation of the hyphal tip. However, unlike the process seen in the *Zygomycetes*, the process here involves something much more like the budding as seen in the yeasts. This is termed a blastic process, which involves the blowing out or blebbing of the hyphal tip wall. The blastic process can involve all wall layers, or there can be a new cell wall synthesized which is extruded from within the old wall. As asexual reproduction is sometimes the only form of reproduction seen in some fungi, we have in the past tried to construct elaborate taxonomic schemes based on spore structure and production. These features are notoriously plastic and such schemes have largely been abandoned. The hypha that creates the sporing (conidiating) tip can be very similar to the normal hyphal tip, or it can be differentiated. The commonest differentiation is the formation of a bottle shaped cell called a phialide, from which the spores are produced.

Not all of these asexual structures are single hyphae. In some groups the conidiophores (the structures that bear the conidia) are aggregated. In the *Moniliales* all are single with the exception of the aggregations termed coremia or synnema. These produce structures rather like corn-stooks, with many conidia being produced in a mass from the aggregated conidiophores.

Other species of *Ascomycetes* and *Deuteromycetes* form their structures within plant tissue, either as parasite and saprophytes. These fungi have evolved more complex asexual sporing structures, probably influenced by the cultural conditions of plant tissue as a substrate. These structures are called the sporodochium. This is a cushion of conidiophores created from a psuedoparenchymatous stroma in plant tissue. The pycnidium is a globose to flask-shaped parenchymatous structure, lined on its inner wall with conidiophores. The acervulus is a flat saucer shaped bed of conidiophores produced under a plant cuticle, which eventually erupt through the cuticle for dispersal.

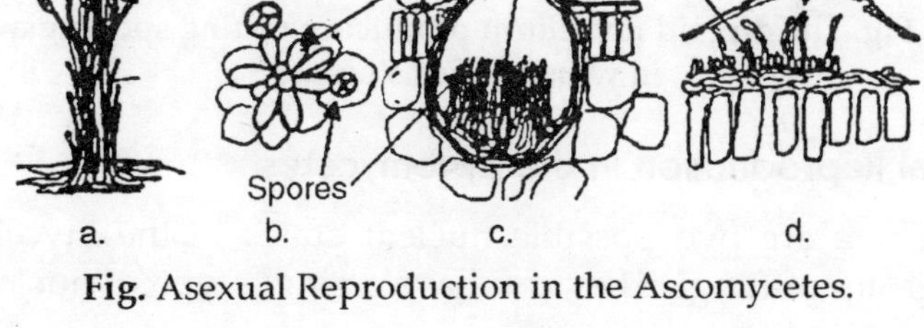

Fig. Asexual Reproduction in the Ascomycetes.

Sexual Reproduction

Sexual reprcduction introduces the possibility of variation into a population, and this is why most fungi have a sexual phase. To achieve sexual reproduction it is necessary to have two mating type haploid nuclei (n + n), or a diploid (2n) nucleus. In the case of the two haploid nuclei they must fuse to form a diploid first, but once fused the nuclei undergo meiosis, which is the reduction division that potentially brings about variation in the progeny. These event are followed by the formation of spores, which in most cases are resting spores that can withstand adverse conditions.

Sexual Reproduction in the Chytridiomycetes

Sexual reproduction occurs in some members of the chytrids by the production of diploid spores after gametic or somatic fusion of two different mating types. The resulting spore may germinate to produce a diploid vegetative mycelium or it may undergo meiosis to produce a haploid mycelium. The diploid mycelium can also produces resting sporangia in which meiosis occurs, generating haploid zoospores that germinate to produce haploid vegetative mycelium:

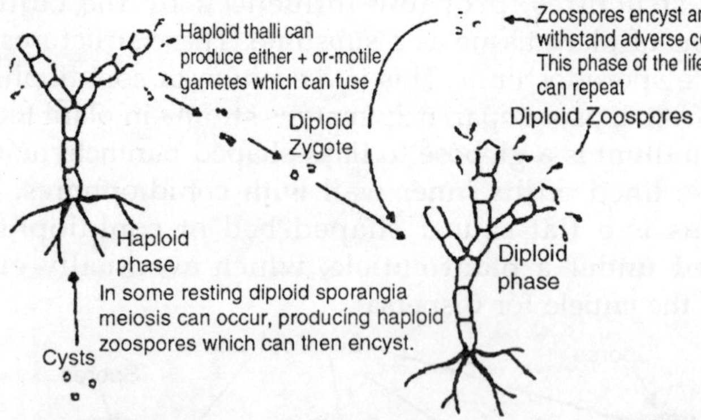

Fig. The diploid mycelium producing resting sporangia in which meiosis occurs.

Sexual Reproduction in the Zygomycetes

There are two possible nuclear states in the mycelia of this group of fungi. They can have a single type of nucleus in their mycelium, a condition termed termed homothallism, or they can contain the two mating type nuclei within their mycelium, termed heterothallism. If the fungus is homothallic the first event in the onset of sexual reproduction has to be somatic fusion. This is termed conjugation. To achieve such a mating it is necessary to attract each other and an elaborate sequence of cellular and biochemical events have been established for some of these fungi. This signalling involves the secretion of inducer molecules that are responsible for causing the formation of zygophores, modified hyphal tips, and these then grow towards each other long a gradient of hormone.

Sexual Reproduction in the Zygomycetes

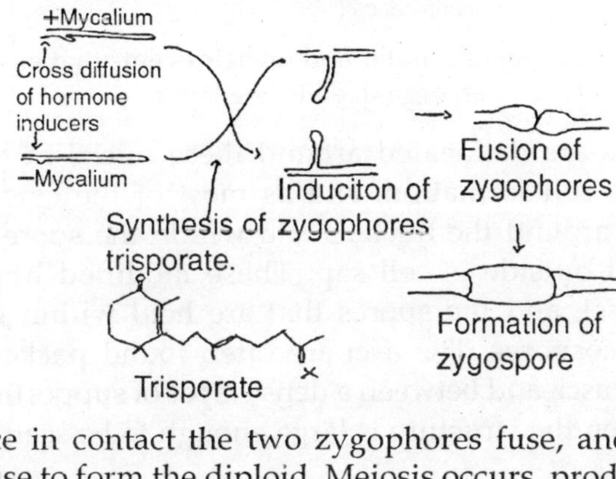

Once in contact the two zygophores fuse, and then the nuclei fuse to form the diploid. Meiosis occurs, producing four haploid nuclei, but three may degenerate. The timing of fusion varies from species to species.

Sexual Reproduction in the Ascomycetes

In this group of fungi there are no specialized organs of hyphal fusion, different mating type mycelia merely fuse with each other to form transient dikaryons, mycelia with two mating type nuclei within it. The dikaryotic mycelium can differentiate to from varying amounts of sterile mycelium around what is to become the fertile tissue of the fruit body. In yeasts, a single, diploid yeast will undergo meiosis, producing four haploid progeny cells, but in more complex fungi there are a sequence of cellular and nucleic events that ensure an organized fertile layer.

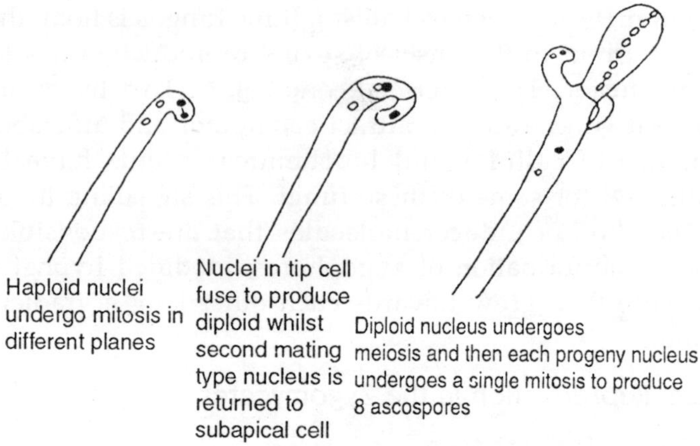

Fig. Sequence of cellular and nucleic events that ensure an organized fertile layer.

Spores are delineated around these nuclei in a process called free cell formation, and as most of the cytoplasm is contained around the nucleus and within the spore wall, all that is left outside is cell sap. These modified hyphae are termed Asci, and the spores that are held within them are termed ascospores. The asci are often found packed tightly with other asci, and between a dense layer of supporting sterile tissue. Often the structure is large enough to be seen with the naked eye.

There are also the yeasts, *Hemiascomycetes*,. Their ascospores are normally formed in loose asci and are not actively discharged. When they form ascospores in fruit pulps or liquids they are usually liberated by the disintegration of the ascus wall.

Sexual Reproduction in the Basidiomycetes

Basidiomycetes are characterised by the most complex and large structures found in the fungi. They are very rarely produce asexual spores. Much of their life cycle is spent as vegetative mycelium, exploiting complex substrates.

A preliminary requisite for the onset of sexual reproduction is the acquisition of two mating types of nuclei by the fusion of compatible mycelium. This creates a dikaryon where single copies of the two mating type nuclei are held within every hyphal compartment for extended periods of time. Maintenance of the dikaryon requires elaborate septum formation during growth and nuclear division.

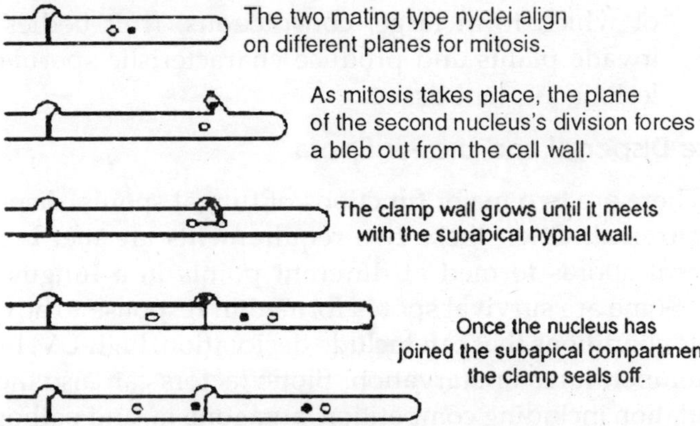

Fig. Reproduction in the Basidiomycetes.

Onset of sexual spore formation is triggered by environmental conditions and in the larger Basidiomycetes begins with the formation of a fruit body primordium. The primordium expands and differentiates to form the large fruit bodies of mushrooms and toadstools. The mycelium within this structure remains as a dikaryon, diploid formation only occurring within the modified hyphal tip called the basidium. Meiosis occurs within the basidium, and the four products are extruded from the tip of the basidium on sterigma. Usually this event occurs across a large area of basidia called a hymenium, or fertile layer. It is usually formed over an extensive sterile layer of tissue like a mushroom gill.

There are three major divisions in the basidiomycetes.

1. *Hymenomycetes*. Basidia are in extensive fertile layer which are susceptible to rain when exposed. Spores are actively discharged from a protected hymenium when ripe. This group includes mushrooms and toadstools, boletes, brackets and coral fungi.

2. *Gasteromycetes.* Hymenia line closed cavities in an initially closed fruit body. Basidiospores are released passively by autolysis of the hymenium, and basidiocarps disintegrate at maturity. This group includes earth balls, puff-balls, stinkhorns and birds nest fungi.
3. *Teliomycetes.* These are the rusts and smuts, neither of which form large, conspicuous, fruit bodies but invade plants and produce characteristic sporulating lesions in plant tissue.

Spore Dispersal and the Air Spora

There are two major functions of fungal spores, dispersal and survival. Often these two requirements are met by two different spores formed at different points in a fungus' life cycle. Some are survival spores formed in response to adverse abiotic conditions that can include desiccation, high UV, high/low temperatures or starvation. Biotic factors can also induce sporulation including competition, antagonism, and pathogens presence. These spores have thick cell walls, and lots of reserves.

Dissemination spores are spores that are smaller, with thin cell walls, and limited reserves, and will germinate readily when on a suitable substrate. They are formed as part of the active life cycle of the fungus and are often concerned with epidemic spread of a pathogenic species from plant to plant, or with rapid colonization of a substrate.

Spores in General

By wet weight spores generally contain 25 per cent protein and 20 per cent fat, and they have a low water content relative to vegetative mycelium. Cell walls of spores are generally not fibrillar, but they are multi-layered and often contain melanin and have ornamentations.

Spores contain all normal mycelial organelles. Respiratory reserves include lipids, glycogen, phospholipids and polysaccharides that can include sugar alcohols like Trehalose).

Respiration rates in spores are only 1-4 per cent those of vegetative mycelium, but obviously the more reserves a spore has, the longer it will survive.

Dormancy

Dormancy occurs when spores do not immediately germinate after formation. Dormancy is a break in the life cycle. There are two types, endogenous (constitutive) and exogenous (induced). Endogenous dormancy is due to some internal quality of the spore, a barrier to water or nutrient entry, a metabolic block, or an inhibitor. Self inhibition prevents spores from germinating in dense suspensions. It can be by excessive sensitivity to oxygen or carbon dioxide levels, nutrient competition, or most usually due to the presence of inhibitors. These molecules are often active in the 1-10 nanomolar range. These inhibitors have to be leached away before germination takes place.

There can also be physical barriers to germination. In one of the athlete foot fungi, *Microsporium gypseum*, there is a protein layer around the spore which prevents the uptake of water. This layer is removed by the action of a fungal acid phosphatase enzyme. This enzyme is inhibited by high levels of phosphate, and until phosphate levels in the environment drop the fungus spore does not germinate.

Endogenous (induced) dormancy occurs because of some external condition, and whilst these conditions prevail the spore will not germinate. As soon as the limiting factor is removed the spore germinates.

Table: A summary of the characteristics of fungal spores with endogenous and exogenous dormancy:

Endogenous dormancy:	Exogenous dormancy:
Displaced from point of origin	Remain at point of origin
Definite launch mechanism	Released by autolysis
Small and thin walled spores	Large and thick walled
Short survival time	Survive for a long time
Germinate readily under suitable conditions	Germinate after a specific stimulus or removal of an inhibitor

Optimal environmental signals trigger the end of dormancy and the onset of germination. Chemical stimuli can trigger germination. This is frequently seen in pathogens where host compounds can act as germination stimulants.

Germination begins with imbibition, the uptake of water, which can cause a 3 to 20 fold increase in size. Spherical growth also accounts for some of the swelling. Eventually polarized growth starts, with the emergence of a germ tube from the spore. The spore wall may be ruptured and a new cell wall covered germ tube emerges, or the spore wall may be softened and the germ tube then emerges.

Air Spora

In between formation of spores and their eventual germination is a phase where spores are disseminated from their point of origin. Many fungi have elaborate mechanisms for getting their spores into the atmosphere which the best medium through which to spread spores. Many spores are very dry and friable, which means they are light enough to be lifted by air currents into the turbulent air above the boundary layer. Others have active spore guns that fire spores up into the atmosphere.

The end result is that over both countryside and towns there is a characteristic air spora. In the countryside this is very typically full of spores from pathogens of agricultural crops, from saprophytes of plant structures and from decaying matter. Species like *Penicillium* and *Cladosporium* tend to predominate. In the towns there are fewer agricultural pathogens but there are still hundreds of spores per cubic litre of air. Within homes and workplaces spore numbers can be even higher, and the species distribution tends to differ. In warm, dry areas *Aspergillus* spp. can become significant members of the air spora.

In normal circumstances these fungal spores pose little or no hazard, our immune systems have evolved with these spores and we are at no risk from them. However, in a population that contains a significant number of individuals that are for whatever reason immuno-compromised, these spores can represent a hazard that as yet has not been quantified. Furthermore, there is increasing evidence to show that inhalation of fungal spores can cause significant allergic responses in atopic individuals. This is an area of very active research.

FUNGI LIFECYCLE

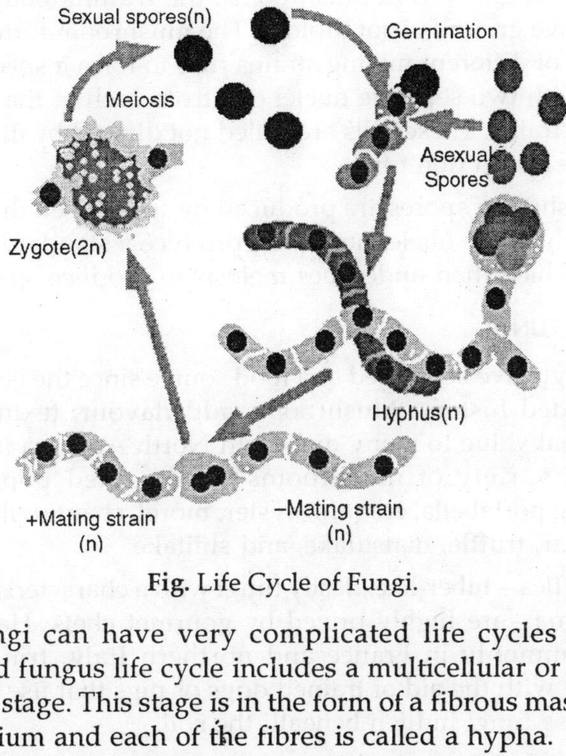

Fig. Life Cycle of Fungi.

Fungi can have very complicated life cycles but the standard fungus life cycle includes a multicellular or colonial haploid stage. This stage is in the form of a fibrous mass called a mycelium and each of the fibres is called a hypha.

Fungus life cycles also include the presence of genetically different mating strains. These look alike but the cells of the hypha. can tell them apart. What's critical for mating is that unlike mating strains (- and+) get together and their nuclei fuse to form the diploid zygote.

The zygote often develops into a resistant thick walled structure often called a zygospore. When conditions are ripe the cell in the zygospore produces sexual sores by meiosis. These are haploid and are dispersed through the air. When they land in a suitable spot such as bread in your bread box, the spores germinated and develop into new hyphae.

Also, the hyphae. will develop specialized structures that produce asexual spores by mitosis.

The key thing is that there is no true diploid multicellular stage in fungi. Even in mushrooms, the fruiting body that is seen above ground is not diploid. The mushroom forms when hyphae. of different mating strains fuse to form a special type of cell with two separate nuclei one from each of the original mating strains. These cells are called not diploid by dikaryotic since the nuclei don't fuse.

Mushroom spores are produced by specialized dikaryotic cell in which the nuclei now fuse, produce a standard diploid zygote which then undergoes meiosis to produce spores.

USES OF FUNGI

Fungi have been used as a food source since the beginning of recorded history. Mushrooms add flavour, texture, and nutritional value to many dishes. In North America in recent years, a variety of mushrooms have gained popularity, including portabella, cremini, oyster, morel, chantarelle, wood or tree ear, truffle, matsutake, and shiitake.

Truffles—tuber-like, fleshy fungi with a characteristic taste and aroma—are highly prized by gourmet chefs. Harvested most commonly in France and northern Italy, truffles are collected with the aid of trained dogs or pigs that use scent to hunt these fungi hidden beneath the soil.

Other fungi are used in the manufacture of foods. Yeast, is added to fruit juice, which it ferments to produce wine. Yeasts also are used in the manufacturing of beer, and they are added to dough to make bread rise, producing more volume and a lighter texture in the final baked product. Certain molds are used to ripen cheeses, such as Brie, Camembert, and the characteristic blue-veined Roquefort. In Asia, fungi are added to soybeans and allowed to ferment to make several

food products—soy sauce is made with the mold *Aspergillus,* and tempeh is made with the black bread mold *Rhizopus.*

Many fungi also produce biologically active compounds that are useful in manufacturing. These compounds include alcohols—such as ethanol and glycerol produced during fermentation—and plant growth regulators—such as giberellic acid, which is used in the promotion of plant and fruit development. Fungi are extremely important in the production of antibiotics; penicillin, griseofulvin, cyclosporine, and cephalosporin are used to fight bacterial and fungal diseases worldwide.

Fungi are becoming an increasingly important tool in cleaning the environment. The accumulation of pesticides and other chemicals in the environment is destroying many ecosystems, and placing many animal and plant species at risk. A number of fungi are used in bioremediation, in which the fungi are mixed with polluted water or soil, where they decompose the organic material in pollutants and, in the process, detoxify them. Fungi employed in this effort include many that are commonly found in soils, such as *Aspergillus, Fusarium, Rhizopus, Mucor, Penicillium,* and *Trichoderma.* Fungi also have been used successfully to control insects, fungus pathogens, roundworms, and other organisms that cause damage and disease to agricultural crops.

HARMFUL FUNGI

Fungi cause about 100,000 diseases of plants, including about 70 per cent of the major crop diseases, resulting in an economic loss of billions of dollars each year. These plant pathogens cause extensive disease to seeds, seedlings, mature plants, and aging plants, resulting in decreased growth and reproduction of crop plants. Fungi also attack forest trees and wooden structures.

A number of fungi cause diseases in humans and other vertebrates. In general, these fungus infections, or mycoses, develop slowly, recur more frequently than bacterial infections, and do not produce a lasting immunity in the body. A mycosis is classified in one of two groups, depending on the part of the body that is infected. A dermatomycosis is an infection of the skin, hair, or nails, such as ringworm or athlete's foot. These infections rarely progress to the internal organs. Most respond well to medication, although treatment may take several weeks.

A systemic mycosis, which is an infection of the entire body, is typically more serious and can be fatal for individuals whose immune system has been weakened by diseases such as acquired immunodeficiency syndrome (AIDS) or cancer. Fungal infections are typically spread by spores that enter the body through inhalation or through an opening in the skin. Some infections are passed from animals to humans or between humans. A few drugs are effective at treating systemic infections, but because treatment may last for several months to years to prevent relapse of the infection, these drugs often cause toxic side effects.

Fungi cause a number of human respiratory diseases. Coccidioidomycosis is caused by the yeast *Coccidioides immitis*. Although typically contracted by the inhalation of dust containing yeast spores, the fungus may also be introduced through the skin from infected soil. Initial symptoms may resemble the flu, with fever, cough, and possibly a skin rash, and the infection usually runs its course without specific treatment. In rare cases, the fungus penetrates internal tissues, such as the bones, joints, and brain, producing tumors that later form abscesses or ulcers. No treatment is available that can halt the course of this form of the disease.

Histoplasmosis is caused by the yeastlike fungus *Histoplasma capsulatum*, which grows in pigeon, bat, and chicken droppings. Contracted by the inhalation of dust from animal droppings, by ingestion, or through the skin. The fungus causing histoplasmosis lives as a parasite in certain tissue and blood cells of the infected person. An infection in the respiratory system is similar to tuberculosis—small spots form in the lungs—although these lesions heal on their own. A progressive form typically invades the bone marrow and is rapidly fatal.

Aspergillosis is an infection of the skin, nasal sinuses, and lungs or other internal organs caused by molds of the genus *Aspergillus*. The disease, contracted by the inhalation of spores, occurs most often among agricultural workers. Itching and pain are frequent symptoms, and if scratching is prolonged, the skin may thicken and become gray or black. A virulent type of pneumonia is caused by the yeastlike fungus *Pneumocystis carinii*, particularly prevalent in people with compromised immune systems, such as AIDS patients.

Mycotoxins are poisons produced by fungal growth in cereals, nuts, fruits, and vegetables. More than 100 species of fungi produce these toxins. The most common mycotoxin is aflatoxin, produced by *Aspergillus flavus* and *Aspergillus parasiticus*. Commonly found on corn, peanuts, and tree nuts, the toxin also can be transmitted to humans through the milk, meat, or eggs of animals fed contaminated grains. Aflatoxin is the most potent carcinogen, or potentially cancer-causing agent, yet discovered. Other mycotoxins include trichothecenes and zearalenone, compounds known to injure the intestines, bone marrow, lymph nodes, spleen, and thymus. They are produced by species of *Fusarium* that grow on grain, straw, or hay stored while damp. Occasionally, circumstances prevent the harvesting of grains during the autumn, and the grains lie dormant in the damp fields until they are harvested in the spring. These grains are especially vulnerable to trichothecenes and zearalenone contamination. A large outbreak of trichothecenes contamination occurred in Russia in early 1944 among hungry peasants who had been searching the winter fields for unharvested wheat and millet.

2

Ascomycetes

The Ascomycota, or sac fungi, is monophyletic and accounts for approximately 75 per cent of all described fungi. It includes most of the fungi that combine with algae to form lichens, and the majority of fungi that lack morphological evidence of sexual reproduction. Among the Ascomycota are some famous fungi: *Saccharomyces cerevisiae*, the yeast of commerce and foundation of the baking and brewing industries (not to mention molecular developmental biology), *Penicillium chrysogenum*, producer of penicillin, *Morchella esculentum*, the edible morel, and *Neurospora crassa*, the "one-gene-one-enzyme" organism. There are also some infamous Ascomycota, a few of the worst being: *Aspergillus flavus*, producer of aflatoxin, the fungal contaminant of nuts and stored grain that is both a toxin and the most potent known natural carcinogen, *Candida albicans*, cause of thrush, diaper rash and vaginitis, and *Cryphonectria parasitica*, responsible for the demise of 4 billion chestnut trees in the eastern USA. Asexual Ascomycota, such as *Penicillium* or *Candida* species, used to be classified separately in the Deuteromycota because sexual characters were necessary for Ascomycota classification. However, the comparison of nucleic acid sequence, as well as nonsexual phenotypic characters, have permitted the integration of asexual fungi into the Ascomycota. The Deuteromycota is no longer recognized as a formal taxon in fungal systematics.

CHARACTERISTICS

The shared derived character that defines the Ascomycota is the ascus. It is within the ascus that nuclear fusion and meiosis take place. In the ascus, one round of mitosis typically follows meiosis to leave eight nuclei, and eventually eight ascospores. Ascospores are formed within the ascus by an enveloping membrane system, which packages each nucleus with its adjacent cytoplasm and provides the site for ascospore wall formation. These membranes apparently are derived from the ascus plasma membrane in the Pezizomycotina and the nuclear membrane in the Saccharomycotina.

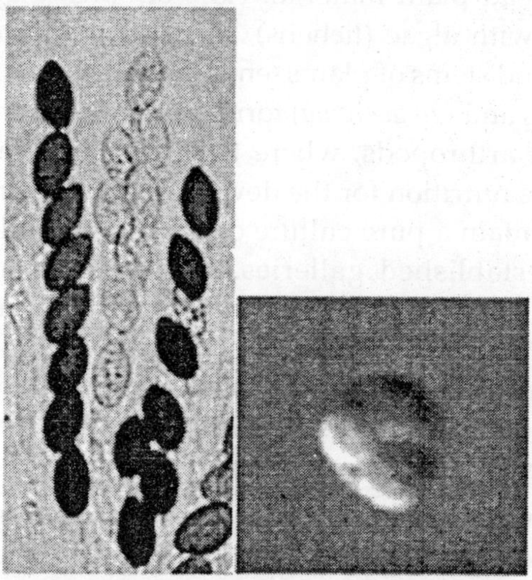

Fig. In hyphal Ascomycota (left), the youngest, terminal hyphal segments develop into 8-spored asci. In yeasts (right) a single cell simply becomes the ascus, often with just 4 spores.

At the time they are released from the ascus, the thick-walled haploid ascospores are resistant to adverse environments. But, given the right conditions, they will germinate to form a new haploid fungus.

The body of Ascomycota is shared by other fungi and consists of a typical eukaryotic cell surrounded by a wall. The body can be a single cell, as in yeasts, or a long tubular filament divided into cellular segments, which is called a hypha. Both yeasts and hyphae have cell walls made of varying proportions of chitin and beta glucans.

NUTRITION AND SYMBIOSES

Like other fungi, Ascomycota are heterotrophs and obtain nutrients from dead or living organisms. If water is present, as saprotrophs they can consume almost any carbonaceous substrate, including jet fuel (*Amorphotheca resinae*) and wall paint (*Aureobasidium pullulans*), and play their biggest role in recycling dead plant material. As biotrophs, they may form symbioses with algae (lichens), plant roots (mycorrhizae) or the leaves and stems of plants (endophytes). Other Ascomycota (*Ceratocystis* and *Ophiostoma*) form symbiotic associations with an array of arthropods, where they can line beetle galleries and provide nutrition for the developing larvae. In return, the beetles maintain a pure culture of the fungus and transport it to newly established galleries. As parasites, ascomycetes account for most of the animal and plant pathogens including *Pneumocystis carinii*, responsible for pneumonia of humans with compromised immune systems and *Ophiostoma ulmi*, the Dutch elm disease fungus that is responsible for the demise of elm trees in North America and Europe.

Ascomycota can be found on all continents and many genera and species display a cosmopolitan distribution (*Candida albicans* or *Aspergillus flavus*). Others are found on more than one continent (*Ophiostoma ulmi*, or *Cryphonectria parasitica*), but many are known from only one narrowly restricted location. The White Piedmont Truffle (*Tuber magnatum*) is known from only one province of Northern Italy.

MODERN CLASSIFICATION OF ASCOMYCOTA

There are 3 subphyla:

1. The subphylum Pezizomycotina.
2. The subphylum Saccharomycotina
3. The subphylum Taphrinomycotina

The Subphylum Pezizomycotina

Pezizomycotina is the largest subphylum of Ascomycota with more than 32,000 described species. It approximately equates to Ascomycetes sensu Kirk et al. 2001 and Euascomycetes sensu Alexopoulos et al. 1996, and it includes all filamentous, sporocarp-producing species, with the exception of *Neolecta* of Taphrinomycotina. Pezizomycotina is ecologically diverse with species functioning in ecological processes and symbioses including wood and litter decay, animal and plant pathogens, mycorrhizae, endophytes and lichens, and occurring in aquatic and terrestrial habitats.

Pezizomycotina includes numerous species that impact humans in both beneficial and detrimental ways. Medically important species have been the source of both disease (e.g., *Coccidioides immitis* – Valley Fever) and life saving drugs (e.g., *Penicillium chrysogenum* – penicillin, *Tolypocladium inflatum* - Cyclosporin A). Introduction of some invasive species have forever altered native forests (e.g., *Cryphonectria parasitica* - Chestnut Blight, *Ophiostoma ulmi* – Dutch Elm Disease), while around 40 per cent of the species form lichens (e.g., Lecanoromycetes) that cover up to 8 per cent of the earth surface and are major sources of fixed carbon in terrestrial environments. Species of Pezizomycotina also have prominent industrial applications including production of organic acids (e.g., *Aspergillus niger* – citric acid) and enzymes and fermented foods (e.g., *Aspergillus oryzae* – amylase, soy sauce).

Characteristics

Most species of Pezizomycotina produce a dominant filamentous or hyphal growth phase, although several species are known to be dimorphic, growing as unicellular budding yeasts under certain conditions. The hyphae of Pezizomycotina are regularly septate with a simple septum that possesses a single pore dividing the hyphae into hyphal compartments or "cells". Associated with the septum of Pezizomycotina is the Woronin body. It is a specialized vesicle that functions to seal the septal pore in response to cellular damage and is comprised of HEX-1 protein that self-assembles to form the solid core of the vesicle. The "cells" of most Pezizomycotina possess a single haploid nucleus, although numerous exceptions to this condition exist.

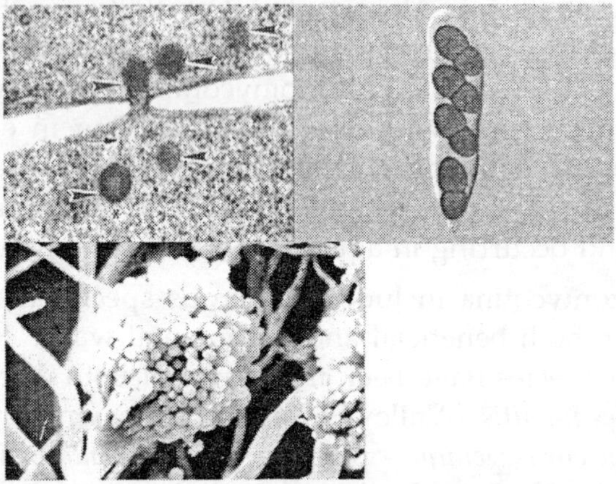

Fig. 1.Hyphal septum with Woronin Bodies. 2. Ascus with ascospores. 3. Mature conidiophores.

As discussed for Ascomycota, the life cycles of Pezizomycotina are highly pleomorphic, with a single species producing 0-1 meiospore-producing stages and 0-many mitospore-producing stages. The meiotic phase (teleomorph) involves the fertilization of a female gametangium (ascogonium) by either a male gametangium (antheridium) or a male gamete (spermatium or microconidium). Fertilized ascogonia develop into dikaryotic ascogenous hyphae, which ultimately develop into asci. Asci are the site of meiosis and meiospore (ascospore) production. The mitotic phase (anamorph) involves the production of mitotic propagules with the most frequent being the production of mitospores (conidia) from specialized conidiogenous cells. Connections between teleomorphs and anamorphs of the same species are frequently unknown as the two often do not co-occur in time (e.g., seasons) and/or space (e.g., host). Thus, as a matter of convenience the International Botanical Code allows unique Latin binomials for both the teleomorph and anamorph with the teleomorph name taking priority. The *Aspergillus nidulans* is the anamorph of the teleomorph *Emericella nidulans*.

Sporocarp and Meiosporangium Diversity

The four main ascomal or sporocarp morphologies include apothecia, perithecia, cleistothecia and ascostromata. Apothecia are typically disc-shaped to cup-shaped to spathulate in morphology and produce asci in a well-defined layer, a hymenium, exposed to the environment. Perithecia and cleistothecia are partially or completely enclosed ascomata, respectively, with ascus production occurring within the central cavity of the ascoma. These three ascomal types are generally presumed to develop after fertilization of the ascogonium (female gametangium). Ascostromata differ in that asci are formed in preformed locules, which often develop in flask shaped (pseudothecia) or open, cup shaped (hysterothecia and thyriothecia) stromatic tissue that superficially resemble perithecia or apothecia, respectively. It is generally presumed that initiation of ascostromata development occurs prior to fertilization of ascogonia.

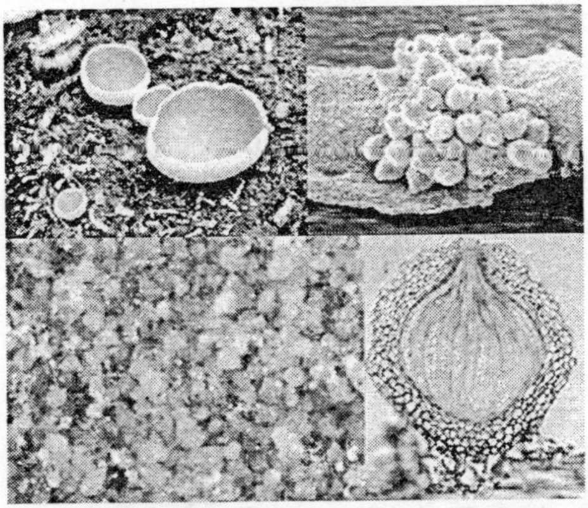

Fig. 1. Apothecia of *Aleuria*. 2. Perithecia of *Neurospora*. 3. Cleistothecia of *Eupenicillium*. 4. Ascostroma of *Tubeufia*.

The major ascus types include operculate, inoperculate, prototunicate, unitunicate and bitunicate, which are based primarily on the number and thickness of functional ascus walls and mechanisms of dehiscence. Operculate asci are restricted to apothecial fungi of Pezizomycetes. They release their ascospores through a defined operculum that is either formed terminally or subterminally at the ascus apex. Inoperculate asci are produced by apothecial (Lecanoromycetes, Leotiomycetes, Orbiliomycetes), cleistothecial (Sordariomycetes, Leotiomycetes) and perithecial (Sordariomycetes) fungi and are typically thin-walled and release their spores through a pore or canal, by rupturing of the ascus apex, or by disintegration of the ascus wall. Prototunicate asci are produced by apothecial (Lecano-romycetes), cleistothecial (Eurotiomycetes) and perithecial (Sordariomycetes) fungi. They are thin-walled, typically globose to broadly clavate, and release their ascospores passively by disintegration of the ascus wall.

Bitunicate asci are conspicuously thick-walled and characterized by possessing two, often separable, functional ascus walls, the exotunica and endotunica. They are produced by ascostromatic lichenized and nonlichenized species (Dothideomycetes, Eurotiomycetes) and ascohymenial lichens (Arthoniomycetes). In the traditional definition of bitunicate asci, ascus dehiscence occurs when the endotunica ruptures through the exotunica in a "jack-in-the-box" manner. Additional modes of dehiscence exist among "bitunicate" ascus morphologies that involve little to no ascus wall separation. Due to the nonfissitunicate mechanisms of dehiscence and relatively thin ascus walls, operculate, inoperculate, and prototunicate asci are collectively referred to as unitunicate.

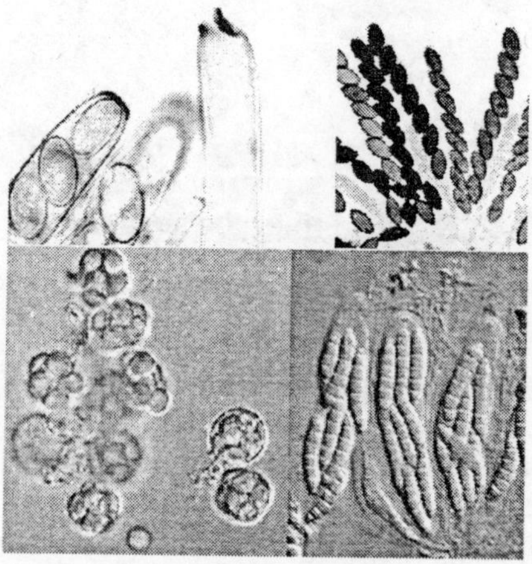

Fig. Operculate asci. 2. Inoperculate asci. 3. Prototunicate asci. 4. Bitunicate asci.

Discussion of Phylogenetic Relationships

Prior to molecular phylogenetics, classifications of sporocarp-producing Ascomycota were based on varying interpretations of morphology and development of ascomata (sporocarps) and asci (meiosporangia). The major taxa and associated character states were: Discomycetes - apothecia, operculate and inoperculate asci; Pyrenomycetes - perithecia, inoperculate asci; Plectomycetes - cleistothecia, prototunicate asci; and Loculoascomycetes - ascostromata, bitunicate asci. Multigene phylogenetic analyses rejected this classification scheme and supported the existence of at least 10 major clades of Pezizomycotina. These studies reveal that apothecia are likely plesiomorphic for the subphylum and that most major ascomal and ascal character states (e.g., cleistothecia and prototunicate asci) have been gained and lost several times.

The two earliest diverging lineages of Pezizomycotina are Orbiliomycetes and Pezizomycetes. Both taxa produce apothecial ascomata but are distinguished by Pezizomycetes producing operculate asci and Orbiliomycetes producing small, inoperculate asci. Current analyses cannot distinguish between either class being the earliest diverging lineage of the Pezizomycotina. The remaining classes of the Pezizomycotina form a well-supported superclass taxon that is informally referred to as 'Leotiomyceta'. Classes of 'Leotiomyceta' include Arthoniomycetes, Dothideomycetes, Eurotiomycetes, Laboulbeniomycetes, Lecanoromycetes, Leotiomycetes, Lichinomycetes, and Sordariomycetes. Superclass relationships among these taxa are mostly unresolved with the exception of the sister group relationships of Arthoniomycetes and Dothideomycetes, and Leotiomycetes and Sordariomycetes, respectively. Geoglossaceae was classified in the Leotiomycetes, but is strongly rejected as a member of the class. It is currently classified 'Leotiomyceta' incertae sedis and may form a clade with Lichinomycetes or represent another class-level lineage.

The Subphylum Saccharomycotina

The subphylum Saccharomycotina comprises most of the "true" yeasts, such as Baker's yeast and *Candida* which are in general single-celled, or short chains of cells, and reproduce vegetatively by budding rather than by the production of hyphae. As a result, most were classified in a vaguely defined taxon with the older name Hemiascomycetes.

The subphylum Taphrinomycotina

The subphylum Taphrinomycotina includes a disparate group of Ascomycota and were only recognized as a distinctive group after the advent of molecular (DNA) analyses. The group is basal to the other subphyla and hence is considered to be more primitive. Consequently the taxon was originally named Archiascomycetes alternatively spelled Archaeascomycetes. It includes both hyphal fungi (*Neolecta, Taphrina*), and fission yeasts *Schizosaccharomyces* and the peculiar mammalian lung parasite, *Pneumocystis* that was originally believed to be a protozoan.

REPRODUCTION

Asexual Reproduction

Asexual reproduction is the dominant form of propagation in the Ascomycota, and is responsible for the rapid expansion of these fungi into areas which were previously not colonized. It occurs through reproductive structures, the "conidia," which are genetically identical to the parent and mostly have just one nucleus. They are also called "mitospores" due to the way they are generated through the cellular process of mitosis. They are generally formed on the ends of specialized hyphae, the "conidiophores". Depending on the species they may be dispersed by wind or water, or also by animals.

Asexual Spores

In order to further classify the Ascomycota in the asexual stages, it is important to consider the spores, which can be distinguished by colour, form and the way they are separated into cells. The most frequent types are the single-celled spores which are designated *amerospores*. If the spore is divided into two by a cross-wall (septum), it is a *didymospore*.

When there are two or more cross-walls the classification depends on the shape. If the septa are *transversal,* like the rungs of a ladder, it is a *phragmospore* whilst if they form a net-like structure it is a *dictyospore.* In *staurospores* ray-like "arms" radiate from a central body; in others (*helicospores*) the entire spore is wound up in a spiral like a spring. Finally very long worm-like spores, of which the ratio length:diameter is more than 15:1, are called *scolecospores.*

Conidiogenesis and Dehiscence

One distinguishes:

- Acervular conidiomata, or *acervuli,* which develop in the host and can thus be:
  a. *Subcuticular,* lying under the outer layer of the plant (the cuticle),
  b. *Intraepidermal,* inside the outer cell layer (the epidermis),
  c. *Subepidermal,* under the epidermis, or
  d. Deeper inside the host.

  Mostly they develop a flat layer of relatively short conidiophores which then produce masses of spores. The increasing pressure finally leads to the splitting of the epidermis and cuticle and so allows the conidia to escape.
- Pycnidial conidiomata or pycnidia, which unlike the acervuli form in the fungal tissue itself, and which are generally shaped like a bulging vase. The spores are released through a small opening at the apex, the ostiole.

Two further important characteristics of the anamorphs of the Ascomycota are the *conidiogenesis,* the fashion in which the spores are formed, and their *dehiscence,* i.e. how they separate from the parent structures. The former corresponds to Embryology in animals and plants and can be divided into two fundamental forms of development: *blastic* conidiogenesis, where the spore is already evident before it separates from the conidiogenic hypha which is giving rise to it, and *thallic* conidiogenesis, where first a cross-wall appears and then the thus created cell develops into a spore.

These two basic types can be further classified as follows.

- Blastic-acropetal (repeated budding at the tip of the conidiogenic hypha, so that a chain of spores is formed with the youngest at the tip),
- Blastic-synchronous (simultaneous spore formation from a central cell, sometimes with secondary acropetal chains forming from the initial spores),
- Blastic-sympodial (repeated sideways spore formation from behind the leading spore, so that the oldest spore is at the main tip),
- Blastic-annellidic (each spore separates and leaves a ring-shaped scar which is concentrically inside the scar left by the previous spore),
- Blastic-phialidic (the spores arise and are ejected from the open ends of special conidiogenic cells called phialides which remain constant in length),
- Basauxic (where a chain of conidia, in successively younger stages of development, is emitted from the mother cell),
- Blastic-retrogressive (spores separate off by formation of crosswalls near the tip of the conidiogenic hypha, which thus becomes progressively shorter),
- Thallic-arthric (double cell walls split the conidiogenic hypha into cells which develop into short, cylindrical spores called arthroconidia; sometimes every second cell dies off, leaving the arthroconidia free),
- Thallic-solitary (a large bulging cell separates from the conidiogenic hypha, forms internal walls, and develops to a *phragmospore*).

Essentially dehiscence can happen in two different ways. In the schizolytic variant a double dividing wall with a central lamella (layer) forms *between* the cells; the central layer dissolves to release the spores. In the case of rhexolytic dehiscence on the other hand the cell wall which joins the spores on the *outside* simply degenerates and sets free the conidia.

Heterocaryosis and Parasexuality

A significant number of Ascomycota species either have no sexual stage or none is known. In spite of this, there are two ways in which they can conserve their genetic diversity: Heterocaryosis and Parasexuality.

The former happens simply through the merging of two hyphae belonging to different individuals, a process known as anastomosis. As a result there are more cell nuclei than normal in the mycelium and they come from genetically different parent organisms.

Parasexuality, on the other hand, refers to a phenomenon where two cell nuclei merge without any sexual process and the chromosome count is doubled. This involves a complex form of the type of cell division called mitosis, where there is *crossing over* or *recombination*, i.e. an exchange of genetic material between corresponding pairs of chromosomes. In sexual reproduction, in contrast, crossing over occurs only during meiosis. Finally the chromosome count will be restored to normal by haploidization, whereby the nucleus splits into two parts each having a single set of chromosomes, with each daughter genetically different from the original parents.

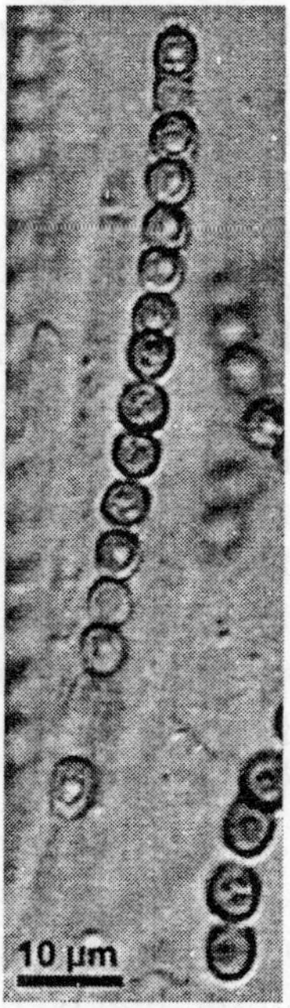

Fig. Ascus of *Hypocrea virens* with eight two-celled Ascospores

Sexual Reproduction

Sexual reproduction in the Ascomycota is marked by a characteristic structure, the *ascus*, which distinguishes these fungi from all others. An ascus is a tube-shaped vessel, a *meiosporangium*, which contains the sexual spores produced by meiosis. The latter are called *ascospores* in contrast to the asexual *conidiospores*.

Apart from exceptions such as Baker's Yeast (Saccharomyces cerevisiae), almost all fungi of the Ascomycota are haploid, so their nuclei only contain one set of chromosomes, which makes them especially susceptible to mutations. During sexual reproduction there is a diploid phase (with two sets of chromosomes), which as a rule is very short. Then meiosis occurs, generally very soon, so that the haploid state is re-established.

The Formation of Sexual Spores

The sexual part of the life cycle commences when two suitable hyphae meet each other. These come from the same web of hyphae which can also generate asexual spores. The first deciding factor as to whether conjugation - that is, sexual merging - will occur, is whether the hyphae belong to the same organism, or whether they come from different individual fungi. Whilst many species are thoroughly capable of self-propagation, i.e. they are homothallic, others need non-identical partners and so are heterothallic. Besides this, the two hyphae in question must also belong to the same mating type. Mating types are a peculiarity of the fungi and correspond roughly to the sexes in plants and animals; however one species may have more than two mating types.

In the case of compatibility, gametangia form on the hyphae; these are the generative cells for the gametes, in which numerous nuclei gather. A very fine hypha, called the trichogyne, which grows out of one gametangium, now termed the ascogonium, makes a passage to a gametangium of the other individual, which is then the antheridium. Nuclei then pass from the antheridium (playing a 'male' role) to the ascogonium (playing a 'female' role).

Unlike the process in animals and plants, after the union of the cytoplasms of the two gametangia (plasmogamy), the merging of the nuclei (karyogamy) does not usually occur immediately. Instead, the nuclei which have migrated in from the antheridium pair up with the nuclei of the ascogonium, but remain separate next to their partners. With this the dikaryophase of the life cycle begins; during this time the pairs of nuclei repeatedly synchronously divide, so that a great number are produced. In all probability the dikaryophase is an evolutionary adaptation which serves to exploit the potential of sexual reproduction to the full in circumstances where it is a rare event for different individuals to meet each other. After the genetic raw material has been increased by repeated division, recombination will take place independently in each pair during meiosis, so that the greatest possible quantity of genetically different spores will arise. In the red algae (Rhodophyta) a similar solution to the corresponding problem evolved independently.

Next millions of new dinucleate hyphae, into each of which two nuclei migrate, emerge from the fertilized ascogonium. They are also called ascogenous or *fertile*. They are fed by ordinary uni- or mononucleate hyphae (with only one nucleus), which are also called *sterile*. The tissue of sterile and fertile hyphae now grows in many cases into a macroscopically visible fruiting body, the ascocarp, which may contain millions of fertile hyphae.

In the actual fruiting layer, the hymenium, the asci now appear. At one end of an ascogenous hypha, there develops a U-shaped hook, which points back opposite to the general growth direction. The two nuclei contained in the terminal cell then divide in such a way that the threads of their mitotic spindles run parallel, and thus two pairs of genetically different daughter nuclei arise, with one daughter of each pair near the point of the hook, and the other in the base part of the hypha. Then two parallel cross-walls appear, dividing the hypha into three sections: that at the point of the hook with one nucleus, that at the base of the original hypha with one nucleus, and the middle U-shaped part with two nuclei.

If the positioning in the fruiting layer is right, the karyogamic fusion of the nuclei finally takes place in the U-shaped cell, creating the diploid zygote. It lengthens to form an elongated tube-shaped or cylinder-shaped capsule, the actual ascus. Then meiosis occurs, giving rise to four haploid nuclei. This is almost always followed by a further mitotic division, so that the ascus ultimately has eight daughter nuclei. These become enclosed, together with some of the cell plasma, each by their own membranes, and generally with a hard cell wall. Thus the dissemination cells (the ascospores) develop, lying initially like peas in a pod inside the ascus. Later, when an appropriate opportunity presents itself, they are liberated.

Not having flagella, ascospores are disseminated in various other ways: some are spread by wind and with others the ripe ascus breaks open on contact with water to set free the spores. Certain species have evolved regular 'spore cannons' which can eject them up to 30 cm. away. When the spores reach a suitable substrate, they germinate, form new hyphae, and so restart their life cycle, which has come full circle.

The form of the ascus is important for classification and is divided into four basic types: unitunicate-operculate, unitunicate-inoperculate, bitunicate, or prototunicate.

LIFE CYCLE

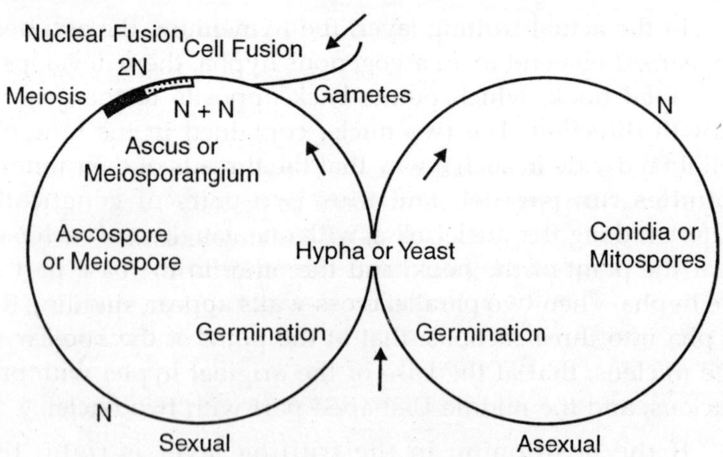

Fig. Life cycle of Ascomycota.

Ascomycota are either single-celled (yeasts) or filamentous (hyphal) or both (dimorphic). Yeasts grow by budding or fission and hyphae grow apically and branch laterally. Most yeasts and filamentous Ascomycota are haploid, but some species, *Saccharomyces cerevisiae*, can also be diploid. Mitospores may simply reproduce the parent, or may also act as gametes to fertilize a compatible partner. Some Ascomycota must outbreed (heterothallic), others can also self, and some can only self (homothallic).

Genetic regulation of sex expression and mating is well-understood in some model Ascomycota such as yeast, where there are two sexes and mating is coordinated by oligopeptide pheromones. In hyphal species, cytoplasmic fusion may not be immediately followed by nuclear fusion, leading to a short dikaryotic phase. The dikaryotic hyphae may be protected and nourished by differentiated haploid hyphae which form a fruiting body (the ascoma; plural ascomata). Ascomata may be closed (cleistothecium), open by a narrow orifice (perithecium), or broadly open like a cup (apothecium). Ascospores are released from the ascoma and germinate to form a new haploid mycelium.

METABOLISM

Like most fungi the Ascomycota principally digest living or dead biomass. To achieve this, they secrete into their surroundings powerful digestive enzymes which break down organic substances into small molecules, which are then absorbed through the cell wall. Many species live on dead plant material such as fallen leaves, twigs, or logs. Others attack plants, animals, or other fungi as parasites and derive their metabolic energy, as well as all the nutrients they need, from the cell tissue of their hosts. Especially in this group extreme specialization appears; for instance certain species of Laboulbeniales attack only one particular leg of one particular insect species. The Ascomycota also often take up symbiotic relationships – for instance some combine with green algae or cyanobacteria, from which they obtain photosynthetic nutrients, to form lichens; others form symbioses with tree roots as mycorrhizal fungi. There are also carnivorous fungi, which have developed hyphal traps in which they can catch small protists such as amoebae, as well as roundworms (*Nematoda*), rotifers, tardigrades, and small arthropods such as springtails (*Collembola*).

Through their long evolutionary history the Ascomycota have developed the capability to break down almost every organic substance. Unlike most organisms they are able to use their own enzymes to digest plant cellulose and the lignin contained in wood. Collagen, an abundant structural protein in animals, and keratin (which hair is made of), can also serve as food sources. Exotic examples are given by the ascomycete *Aureobasidium pullulans*, which metabolizes wall paint, and the kerosene fungus *Amorphotheca resinae*, which (to the misfortune of the airline industry) feeds on aircraft fuel, and in tropical regions sometimes blocks fuel pipes. Others resist osmotic stress to grow on salted fish, and a few live in water.

Relationships of Ascomycota to other Fungi

The Ascomycota is sister group to the Basidiomycota. This relationship is supported by the presence in members of both phyla of cross-walls (septa) that divide the hypahe into segments, and pairs of unfused nuclei in these segments after mating and before nuclear fusion (dikaryons). Further support comes from the apparent homology between structures that coordinate simultaneous mitosis of the two dikaryotic nucli (Ascomycota croziers and Basidiomycota clamp-connections).

Discussion of Phylogenetic Relationships

Molecular phylogenetic analyses of nuclear and mitochondrial ribosomal RNA genes and protein coding genes support a monophyletic Ascomycota. Early diverging lineages of Ascomycota have been classified in Taphrinomycotina. Due to the lack of strong support for the monophly of Taphrinomycotina and the paraphyletic resolution of these taxa in some analyses, Taphrinamycotina is not recognized in some classifications. More recent analyses, however, that included multiple protein coding genes and RNA genes recovered a monophyletic Taphrinomycotina with greater support.

The Taphrinomycotina includes yeast species (*Pneumocystis, Schizosaccharomyces*), dimorphic taxa (*Taphrina* spp.) and a filamentous sporocarp producing genus (*Neolecta*).

The placement of *Neolecta* among the basal lineages of the Ascomycota is surprising because of the presence of an ascoma, a feature not found in the other basal lineages or in any Saccharomycotina. However, there is no reason that the Saccharomycotina could not have lost ascomata as hyphal growth became suppressed in favour of yeasts. The Saccharomycotina form a well-supported monophyletic taxon, as do the Pezizomycotina. Asexual fungi sharing morphological or molecular characters of sexual Ascomycota are classified in the Ascomycota and its subtaxa include *Candida albicans* (Saccharomycotina, Saccharomycetes) and *Pencillium chrysogenum* (Pezizomycotina, Eurotiomycetes).

By comparing nucleic acid sequences from 50 genes, the timing of Ascomycota evolution has been estimated, although results produced a wide geological time span depending on calibrations points used. The Taphrinomycotina, Saccharomycotina and Pezizomycotina were likely established in the early Devonian, a bit more than 400 million years ago (mya). Some estimates, however, suggest a much earlier Ascomycota origin of ca. 1000 mya. Fossils of early Ascomycota are not easy to recognize and the utility of some of them as exemplars of extant lineages is problematical (e.g., *Paleopyrenomycites devonicus* as a fossil Sordariomycetes). Thus, we still rely on generally accepted fossil dates external to Fungi (e.g., dicot-monocot split) for potentially more robust calibration points.

Subgroups of Ascomycota

The basal lineage or lineages of Ascomycota comprise four classes, Neolectomycetes, Pneumocystidomycetes, Schizosaccharomycetes, and Taphrinomycetes, which are classified in the subphylum, Taphrinomycotina (=Archiascomycetes Nishida and Sugiyama). The monophyly of Taphrinomycotina is questionable, but increased sampling of both taxa and genes has resulted in increased support for the monophyly of the taxon.

The Saccharomycotina comprises the 'true yeasts' and is home to the most famous fungus, *Saccharomyces cerevisiae*, better known as the baker's yeast. Although most members are primarily unicellular, the basal taxa make abundant hyphae. The Saccharomycotina lack ascomata. Current classification of the Saccharomycotina includes one class, Saccharomycetes, and one order, Saccharomycetales.

Pezizomycotina contain well over 90 per cent of Ascomycota, and the species are hyphal, with almost all of the sexually reproducing forms possessing ascomata. The Pezizomycotina includes 11 well supported clades that are recognized as classes. Most of the recent molecular phylogenetic effort has been directed at this subphylum.

ECOLOGY

The Ascomycota fulfil a central role in most land-based ecosystems. They are important decomposers which break down such organic materials as dead leaves, twigs, fallen trees, etc. and help the detritivores (animals which live off this decomposing material) to obtain their nutrients. By processing substances like cellulose or lignin, which are otherwise difficult to exploit, they take on an important place in the natural nitrogen cycle and the carbon cycle.

Inversely the fruiting bodies of the Ascomycota provide food for a very diverse set of animals from insects and slugs and snails (*Gastropoda*) to rodents and larger mammals such as deer and wild boars.

Fungi of the Ascomycota are also known for their numerous symbiotic relationships with other organisms.

Lichens

Probably since early in their evolutionary history the Ascomycota have "domesticated" green algae (*Chlorophyta*), as well as occasionally other types of algae and cyanobacteria. Together they form the mutualistic associations known as lichens, which can survive in the least hospitable regions of the earth, including the Arctic, the Antarctic, deserts and mountaintops, and can withstand temperature extremes from -40°C to +80°C. While the photoautotrophic algal partner creates metabolic energy through photosynthesis, the fungus offers a stable supportive framework and protects from radiation and drying out. Around 42 per cent of the Ascomycota (numerically about 18,000 species) form lichens, and almost all the fungal partners of lichens belong to the Ascomycota - the proportion of Basidiomycota is probably only two to three per cent.

Mycorrhizal Fungi and Endophytes

Members of the Ascomycota make two particularly important types of relationship with plants: as mycorrhizal fungi and as endophytes. The former make symbiotic associations with the root systems of the plants, which for some trees, especially conifers, can be of vital importance, enabling the uptake of mineral salts from the soil. The fungal partner is in a much better position to absorb minerals due to its finely divided mycelium, whilst the plant provides it with metabolic energy in the form of photosynthetic products. Cases are even known where mycorrhizal fungi can transport nutrients from one plant to another, stabilizing the recipient. It is likely that mycorrhizal associations enabled the conquest of the land by plants - in any case the earliest known fossils of land plants have mycorrhizae.

Endophytes on the other hand live inside plants, especially in the stem and leaves, but generally do not damage their hosts. The exact nature of the relationship between endophytic fungus and host is not yet well understood, but it seems that this form of colonization can bestow a higher resistance against insects, roundworms (nematodes), and bacteria; also it can enable or augment the production of poisonous alkaloids, chemicals which can affect the health of plant-eating mammals.

Symbiotic Relationships with Animals

A series of Ascomycota species from the genus *Xylaria* are found in the nests of leafcutter ants and other fungus-growing ants of the tribe *Attini* and in the fungal gardens of termites (*Isoptera*). Since they do not generate fruiting bodies until the insects have left the nests, it is suspected that, as confirmed in several cases of Basidiomycota species, they may be cultivated.

On the other hand bark beetles (*Scolytidae*) are certainly important symbiotic partners. The female beetles transport the spores to new hosts in characteristic tucks in their skin, the *mycetangia*. There they eat tunnels in the wood, which lead into large chambers in which they lay their eggs. At this time the spores are released and give rise to hyphae which unlike the beetles can digest the wood. The beetle larvae feed on the fungus and after they have metamorphosed into the adult state they again carry spores with them to renew the cycle of infection. Dutch elm disease, caused by fungus *Ophiostoma ulmi*, being carried by the European elm bark beetle *Scolytus multistriatus*.

3

Basidiomycetes

The Basidiomycota contains about 30,000 described species, which is 37 per cent of the described species of true Fungi. The most conspicuous and familiar Basidiomycota are those that produce mushrooms, which are sexual reproductive structures. The Basidiomycota also includes yeasts and asexual species. Basidiomycota are found in virtually all terrestrial ecosystems, as well as freshwater and marine habitats.

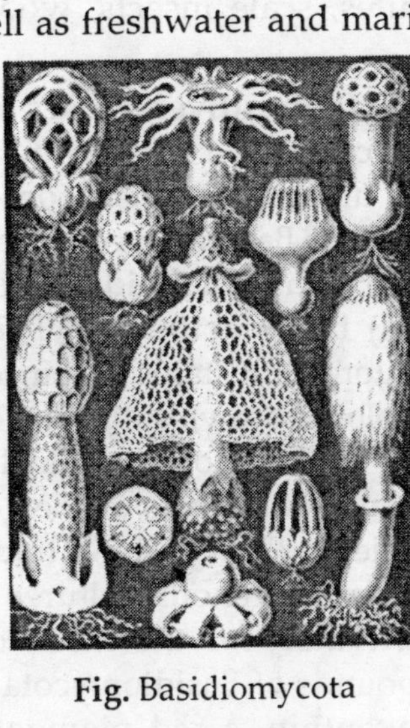

Fig. Basidiomycota

Basidiomycota have a huge impact on human affairs and ecosystem functioning. Many Basidiomycota obtain nutrition by decaying dead organic matter, including wood and leaf litter. Thus, Basidiomycota play a significant role in the carbon cycle. Unfortunately, Basidiomycota frequently attack the wood in buildings and other structures, which has negative economic consequences for humans.

Symbiotic lifestyles (intimate associations with other living organisms) are well developed in the Basidiomycota. Symbiotic Basidiomycota include important plant pathogens, such as "rusts" (Uredinales) and "smuts" (Ustilaginales), which attack wheat and other crops. Other symbiotic Basidiomycota cause diseases in animals, including humans. Not all symbiotic Basidiomycota cause obvious harm to their partners, however. Some Basidiomycota, as well as a handful of Ascomycota, form ectomycorrhizae, which are associations with the roots of vascular plants (principally forest trees such as oaks, pines, dipterocarps, and eucalypts). Ectomycorrhizal Basidiomycota help their plant partners obtain mineral nutrients from the soil, and in return they receive sugars that the plants produce through photosynthesis. Other symbiotic Basidiomycota form associations with insects, including leaf-cutter ants, termites, scale insects, woodwasps, and bark beetles.

Humans have found diverse uses for Basidiomycota. Mushrooms, both cultivated and wild, are eaten in many countries. For the untrained, mushroom-hunting is a risky endeavor, because some Basidiomycota produce deadly toxins. The basidiomycete toxin phalloidin (from the mushroom *Amanita phalloides*) binds actin, which is a component of microfilaments. Fluorescent stains that incorporate phalloidin are used by cell biologists to visualize the cytoskeleton. Other "toxins" produced by Basidiomycota include hallucinogens, which are produced by members of the genus *Psilocybe* (and other groups). Species of *Psilocybe* have traditionally been used in Central American indigenous cultures as a spiritual tool, and are now cultivated for the illicit drug trade. Other biochemical compounds of Basidiomycota that have practical uses include astaxanthin, a red pigment produced by the basidiomycetous yeast *Phaffia* (used to add colour to farmed salmon), and certain enzymes from wood-decaying

Basidiomycota that have potential applications in paper production and bioremediation (decontamination of polluted environments using biological agents).

CHARACTERISTICS

Basidiomycota are unicellular or multicellular, sexual or asexual, and terrestrial or aquatic. Indeed, Basidiomycota are so variable that it is impossible to identify any morphological characteristics that are both unique to the group and constant in the group. The most diagnostic feature is the production of basidia (sing. basidium), which are the cells on which sexual spores are produced, and from which the group takes its name. A long-lived dikaryon, in which each cell in the thallus contains two haploid nuclei resulting from a mating event, is another characteristic feature. Finally, clamp connections are a kind of hyphal outgrowth that is unique to Basidiomycota, although they are not present in all Basidiomycota. The description of the characteristics of Basidiomycota traces the life cycle of a "typical" species, beginning at the site of meiosis.

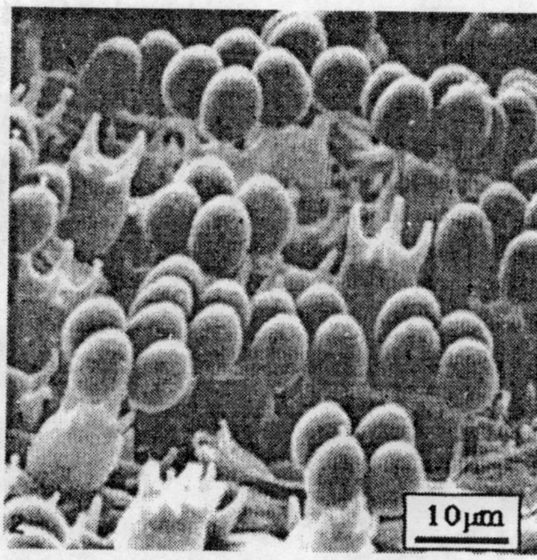

Fig. SEM of the surface of a mushroom gill showing several basidia, some with four basidiospores attached.

The basidium is the cell in which karyogamy (nuclear fusion) and meiosis occur, and on which haploid basidiospores are formed (basidia are not produced by asexual Basidiomycota). Many Basidiomycota produce basidia on multicellular fruiting bodies (e.g., mushrooms), but basidia can also be formed directly from yeasts or other single cells. There is a great range of variation in morphology of the basidium, the number of spores formed, and how the spores are borne on the surface of the basidium. Typically, four spores are produced on each basidium, at the tips of minute stalks called sterigmata. Each spore usually contains one or two of the haploid meiotic products.

One of the most fascinating characteristics of Basidiomycota is the production of forcibly discharged ballistospores, which are propelled into the air from the sterigma. Ballistospores may be sexual or asexual, and may be produced by basidia, hyphae, yeast cells, or even other ballistospores. This type of spore discharge must have evolved very early in the evolutionary history of the Basidiomycota as it is found in members of the earliest diverging lineages within the group. Ballistospory is associated with forms that disperse their spores directly into the air. Most aquatic Basidiomycota and forms that produce spores inside the fruiting body, such as puffballs, have lost ballistospory.

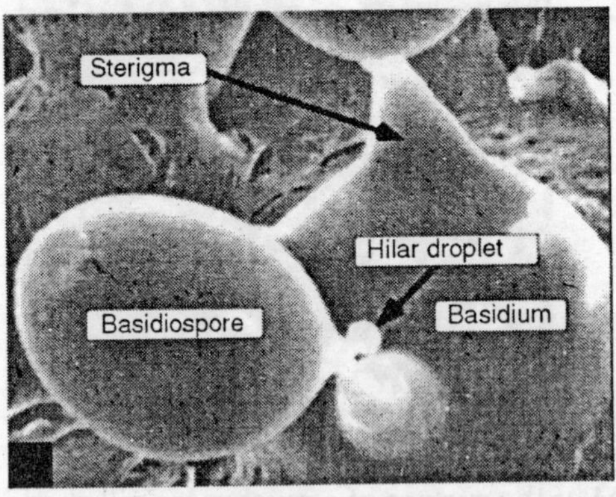

Fig. Scanning electron micrograph showing the hilar droplet at the base of a basidiospore.

Ballistospory is associated with the production of a liquid filled "hilar droplet" that forms at the base of the spore, just above its attachment to the sterigma. Resolving the mechanism of ballistospory has been a longstanding problem in mycology.

In a series of studies, reviewed by Money, it has been shown that spore discharge occurs when the hilar droplet fuses with a film of liquid on the surface of the spore. The rapid coalescence of the liquids causes a sudden shift in the centre of mass of the spore and contributes to its release from the sterigma. This mechanism has been termed a "surface tension catapult" and it results in spores being discharged with a force of about 25,000 *g*.

Basidiospores germinate to form hyphae (filaments) or yeast cells that are typically haploid and uninucleate. The hyphae of Basidiomycota are septate. Ultrastructural features of the septa, visible with transmission electron microscopy, have been important in developing phylogenetic hypotheses in Basidiomycota.

Mating in Basidiomycota involves fusion of haploid cells, but fusion of the nuclei is usually delayed until the basidia are formed. Thus, the dominant phase of the life cycle in most Basidiomycota is a dikaryon, in which the two nuclei brought together in mating exist side-by-side in each cell. Sometimes a dikaryon can donate a nucleus to a uninucleate "monokaryon", resulting in a "di-mon" mating.

Clamp connections are hyphal outgrowths that form when cells in dikaryotic hyphae divide. One of the nuclei divides in the main axis of the hypha, while the other divides into the clamp. Septa are formed across each of the mitotic spindles. The apex of the backward-growing monokaryotic clamp cell fuses with the subapical cell, reestablishing the dikaryotic condition. All fungi that produce clamp connections are members of the Basidiomycota, but not all Basidiomycota produce clamp connections. The regular formation of clamp connections must have developed early in basidiomycete evolution, because they are found in all the major clades of Basidiomycota.

Sexually reproducing Ascomycota also form dikaryons, although they are not as long-lived as those of Basidiomycota. The clade that includes Ascomycota and Basidiomycota has been called the Dikaryomycotina, reflecting this presumably homologous similarity. Ascomycota produce clamp-like "croziers" at the bases of asci (cells in which meiosis occurs, homologous to basidia). Croziers may be homologous to clamp connections.

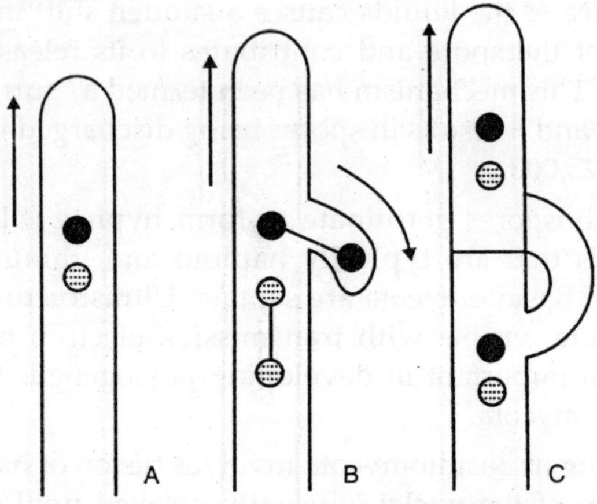

Fig. Diagram of clamp cell formation: A. Dikaryotic hypha, arrow shows direction of hyphal tip growth. B. Clamp cell growing backward, nuclei undergoing synchronous division. C. Mature clamp.

TYPICAL LIFE-CYCLE

Unlike higher animals and plants which have readily recognizable male and female counterparts, Basidiomycota (except for the Rust (Pucciniales)) tend to have mutually indistinguishable, compatible haploids which are usually mycelia being composed of filamentous hyphae. Typically haploid Basidiomycota mycelia fuse via plasmogamy and then the compatible nuclei migrate into each others mycelia and pair up with the resident nuclei. Karyogamy is delayed, so that the compatible nuclei remain in pairs, called a dikaryon. The hyphae are then said to be dikaryotic. Conversely, the haploid mycelia are called monokaryons. Often, the dikaryotic mycelium is more vigourous than the individual monokaryotic mycelia, and usually the dikaryotic mycelium then takes over the substrate in which they are growing. The dikaryons can be long-lived, lasting years, decades, or centuries. *The monokaryons are neither male nor female.* They have either a bipolar (unifactorial) or a tetrapolar (bifactorial) mating system. This results in the fact that following meiosis, the resulting haploid basidiospores and resultant monokaryons, have nuclei that are compatible with 50 per cent (if bipolar) or 25 per cent (if tetrapolar) of their sister basidiospores (and their resultant monokaryons) because the mating genes must differ for them to be compatible. However, there are many variations of these genes in the population, and therefore, over 90 per cent of monokaryons are compatible with each other. It is as if there were multiple sexes.

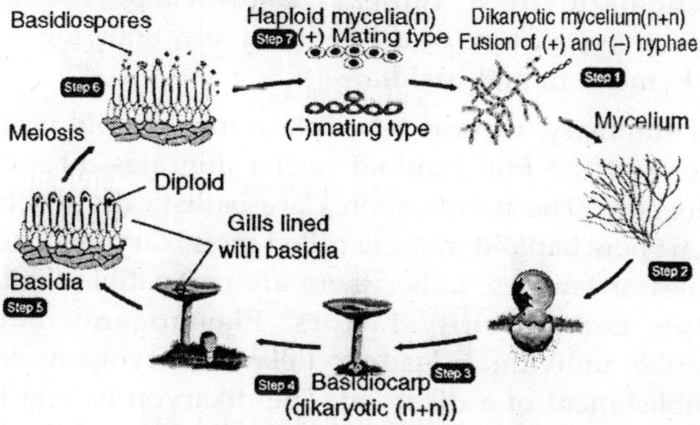

Fig. Life cycle of Basidiomycetes

The maintenance of the dikaryotic status in dikaryons in many Basidiomycota is facilitated by the formation of clamp connections that physically appear to help coordinate and re-establish pairs of compatible nuclei following synchronous mitotic nuclear divisions. Variations are frequent and multiple. In a typical Basidiomycota lifecycle the long lasting dikaryons periodically (seasonally or occasionally) produce basidia, the specialized usually club-shaped end cells, in which a pair of compatible nuclei fuse (karyogamy) to form a diploid cell. Meiosis follows shortly with the production of 4 haploid nuclei that migrate into 4 external, usually apical basidiospores. Variations occur, however. Typically the basidiospores are ballistic, hence they are sometimes also called ballistospores. In most species, the basidiospores disperse and each can start a new haploid mycelium, continuing the lifecycle. Basidia are microscopic but they are often produced on or in multicelled large fructifications called basidiocarps or basidiomes, or fruitbodies), variously called mushrooms, puffballs, etc. Ballistic basidiospores are formed on sterigmata which are tapered spine-like projections on basidia, and are typically curved, like the horns of a bull. In some Basidiomycota the spores are not ballistic, and the sterigmata may be straight, reduced to stubbs, or absent. The basidiospores of these non-ballistosporic basidia may either bud off, or be released via dissolution or disintegration of the basidia.

Schematic of a typical basidiocarp, the dipoid reproductive structure of a basidiomycete, showing fruiting body, hymenium and basidia.

In summary, meiosis takes place in a diploid basidium. Each one of the four haploid nuclei migrates into its own basidiospore. The basidiospores are ballistically discharged and start new haploid mycelia called monokaryons. There are no males or females, rather there are compatible thalli with multiple compatibility factors. Plasmogamy between compatible individuals leads to delayed karyogamy leading to establishment of a dikaryon. The dikaryon is long lasting but ultimately gives rise to either fruitbodies with basidia or directly to basidia without fruitbodies. The paired dikaryon in the basidium fuse (i.e karyogamy takes place). The diploid basidium begins the cycle again.

Variations in Life-cycles

Many variations occur. Some are self compatible and spontaneously form dikaryons without a separate compatible thallus being involved. These fungi are said to be *homothallic*, versus the normal *heterothallic* species with mating types. Others are secondarily homothallic, in that two compatible nuclei following meiosis migrate into each basidiospore, which is then dispersed as a pre-existing dikaryon. Often such species form only two spores per basidium, but that too varies. Following meiosis, mitotic divisions can occur in the basidium.

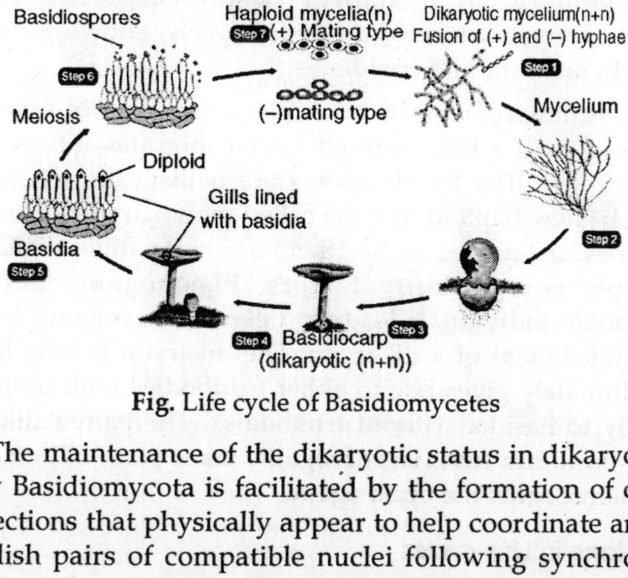

Multiple numbers of basidiospores can result, including odd numbers via degeneration of nuclei, or pairing up of nuclei, or lack of migration of nuclei. The chanterelle genus *Craterellus* often has 6-spored basidia, while some corticioid *Sistotrema* species can have 2-, 4-, 6-, or 8-spored basidia, and the cultivated button mushroom, *Agaricus bisporus*. can have 1-, 2-, 3- or 4-spored basidia under some circumstances. Occasionally monokaryons of some taxa can form morphologically fully formed basidiomes and anatomically correct basidia and ballistic basidiospores in the absence of dikaryon formation, diploid nuclei, and meiosis. A rare few number of taxa have extended diploid life-cycles, but can be common species. Examples exist in the mushroom genera *Armillaria* and *Xerula*, both in the Physalacriaceae. Occasionally basidiospores are not formed and parts of the "basidia" act as the dispersal agents, e.g. the peculiar mycoparasitic jelly fungus, *Tetragoniomyces* or the entire "basidium" acts as a "spore", e.g. in some false puffballs (*Scleroderma*). In the human pathogenic genus *Filobasidiella* 4 nuclei following meiosis remain in the basidium but continually divide mitotically, each nucleus migrating into synchronously forming nonballistic basidiospores that are then pushed upwards by another set forming below them, resulting in 4 parallel chains of dry "basidiospores".

Other variations occur, some as standard life-cycles (that themselves have variations within variations) within specific orders.

Phylogenetic Relationships

There is strong evidence that the Basidiomycota is monophyletic. Ballistospores, basidia, and clamp connections are present in the Agaricomycotina, Ustilaginomycotina, and Pucciniomycotina (although not in all species), suggesting that they have a common origin. Non-molecular characters that have been used to recognize major groups within the Basidiomycota include the form of the basidia (shape and septation), ultrastructure of hyphal septa and spindle pole bodies, presence or absence of yeast phases and "spore repetition" (production of spores directly from spores), and cellular carbohydrate composition. Sequences of ribosomal RNA (rRNA) genes, recently supplemented by protein-coding genes, have played a major role in increasing our understanding of the relationships within Basidiomycota, and have demonstrated that some morphological attributes that have been emphasized in higher-level classification, such as the form of basidia, are subject to homoplasy.

CLASSIFICATION OF BASIDIOMYCETES

The most recent classification adopted by a coalition of 67 mycologists recognizes 3 subphyla (Pucciniomycotina, Ustilaginomycotina, Agaricomycotina) and 2 other class level taxa (Wallemiomycetes, Entorrhizomycetes) outside of these, among the Basidiomycota. As now classified, the subphyla join and also cut across various obsolete taxonomic groups previously commonly used to describe various Basidiomycota.

Agaricomycotina

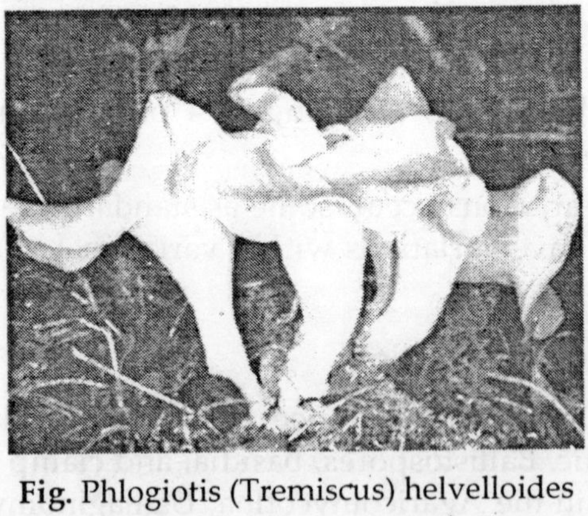

Fig. Phlogiotis (Tremiscus) helvelloides

The Agaricomycotina is one of three major clades of Basidiomycota (the others are the Pucciniomycotina and Ustilaginomycotina). The Agaricomycotina contains roughly 20,000 described species, which is almost 70 per cent of the (known) Basidiomycota. About 98 per cent of the species of the Agaricomycotina are in a clade called the Agaricomycetes, which includes mushrooms, bracket fungi, puffballs, and others. The other major groups are the Tremellomycetes and Dacrymycetes. These latter groups include "jelly fungi", which have gelatinous, often translucent fruiting bodies (e.g., "witches butter" *Tremella mesenterica*), as well as many yeast-forming species.

Members of the Agaricomycotina display the full range of ecological strategies that characterizes the Basidiomycota as a whole. To obtain carbon nutrition, Agaricomycotina decompose dead organic matter or enter into diverse associations (both antagonistic and benign) with plants, animals, and other fungi. Mycorrhizal associations with plants are present in many lineages of Agaricomycetes, including certain jelly fungi. Mycoparasitism (parasitism of a fungus by a fungus) is especially widespread in the Tremellomycetes. Few Agaricomycotina are medically important. An exception is *Filobasidiella neoformans* (also called *Cryptococcus neoformans*), which is a serious pathogen of immunocompromised individuals.

Agaricomycotina include the vast majority of edible mushrooms, most of which are in the Agaricomycetes. Several jelly fungi are cultivated for food in Asia, including the "wood ear" fungus, *Auricularia auricula-judae*, and the white jelly fungus, *Tremella fuciformis*.

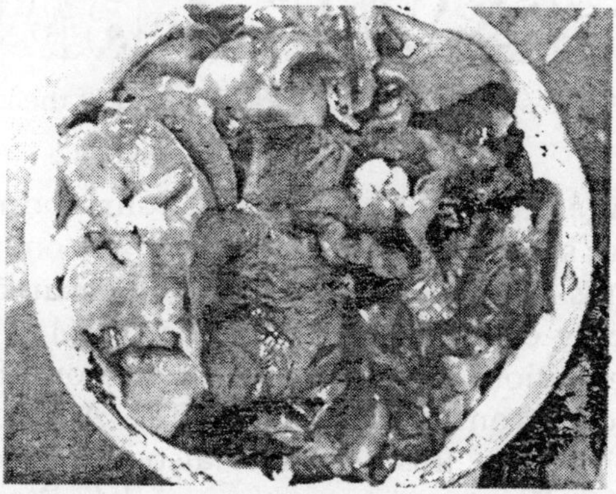

Fig. Cultivated fruiting bodies of *Auricularia auricula-judae*

Characteristics

Agaricomycotina are quite variable in both macroscopic and microscopic features. Traditionally, one of the most important characters in the higher-level taxonomy of Basidiomycota has been the form and septation of the basidia (the cells in which meiosis occurs, and on which sexual spores are formed). Basidia of Agaricomycotina have various shapes and may be undivided or divided by transverse or longitudinal septa. Previously—including in a former version of this Tree of Life page—the Agaricomycotina with "homobasidia" (which are undivided basidia that produce spores on short apical appendages, called sterigmata) were classified as the "Homobasidiomycetes". However, recent phylogenetic studies suggest that clades with undivided homobasidia are interdigitated with clades that have septate "heterobasidia". Consequently, the terms "homobasi-diomycetes" and "heterobasidiomycetes" are no longer used to refer to formal taxa, but rather are used as descriptive terms only.

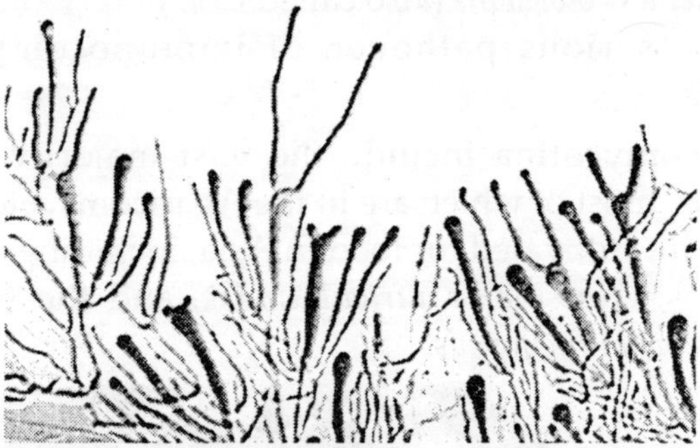

Fig. Two-spored "tuning fork" basidia of the jelly fungus *Dacrymyces*.

Most Agaricomycotina produce four spores on each basidium, but some species produce as few as one or as many as eight spores per basidium. Another important character in prior taxonomy of Basidiomycota is the presence or absence of "spore repetition", which is the production of secondary spores directly from basidiospores. The Agaricomycotina display variation in this character also. It is therefore not surprising that the Agaricomycotina, as delimited here, was not formally recognized as a taxonomic group before the advent of molecular characters.

One unifying morphological feature of the Agaricomycotina is the presence of "dolipore" septa, which are pores in the crosswalls between adjacent cells that are bordered by a collar-like margin of cell wall material. Other Basidiomycota also have septal pores, but their structure is not the same. Flanking the dolipores in almost all Agaricomycotina is a membrane-bound structure called a "septal pore cap". These structures are also called "parenthesomes", because they resemble parentheses around the septal pores when viewed in thin sections with transmission electron microscopy. Variation in the form of the parenthesomes has provided clues to phylogenetic relationships in the Agaricomycotina. Tremellomycetes have "vesiculate" parenthesomes, in which the septal pore cap is divided into cup-shaped sections. Dacrymycetes and some Agaricomycetes have "imperforate" parenthesomes, in which the septal pore cap forms a smooth, dome-shaped covering over the dolipore. Most Agaricomycetes have "perforate" parenthesomes, in which the septal pore cap has openings of varying sizes, and appears discontinuous in section.

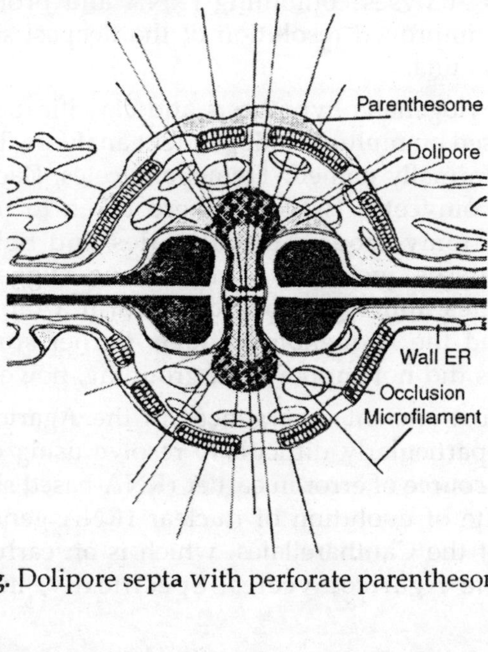

Fig. Dolipore septa with perforate parenthesome.

Most known species of Agaricomycotina are filamentous and produce a multicellular fruiting body. The fruiting body is not always what most people would regard as a "mushroom", however. In "resupinate" species the fruiting body is a flattened structure, which may be little more than a thin layer of hyphae produced on the underside of a log or some other inconspicuous location. As noted, some Agaricomycotina are yeasts and produce no fruiting body at all. Fruiting bodies of Agaricomycotina reach their greatest size, diversity, and complexity in the Agaricomycetes. It is likely that the yeasts and resupinate forms have been undercollected, relative to their more conspicuous relatives.

Phylogenetic Relationships

As in all groups of Fungi, understanding of phylogenetic relationships in Agaricomycotina has advanced rapidly in recent years. Early phylogenetic analyses using rRNA gene sequences resolved many major groups within the Hymenomycetes, but higher-level relationships among these groups were often not resolved with confidence. Recently, multi-locus analyses combining rRNA and protein-coding genes have improved resolution of the deepest splits in the Agaricomycotina.

The phylogenetic hypothesis and classification adopted here are based on multiple independent analyses. Three major clades are formally named: Tremellomycetes, Dacrymycetes, and Agaricomycetes (which is here taken to include the heterobasidiomycetous Auriculariales and Sebacinales). Several analyses, including a multi-locus, suggest that the Dacrymycetes and Agaricomycetes form a monophyletic group called the Hymenomycetidae. Earlier studies using rRNA genes did not support that grouping, however.

The most ancient divergences in the Agaricomycotina have been particularly difficult to resolve using only rRNA genes. One source of error in earlier rRNA-based studies is an elevated rate of evolution of nuclear rRNA genes in some members of the Cantharellales, which is an early-diverging clade in the Agaricomycetes. Specifically, members of

Cantharellus and *Craterellus* (which include edible chanterelles) and *Tulasnella* (a group of resupinate "crust fungi") have highly divergent nuc rRNA sequences that cannot be aligned in their entirety to other Agaricomycetes and produce conspicuously long branches in phylograms. Analyses using protein-coding genes leave little doubt that these taxa are in the Cantharellales clade of the Agaricomycetes.

USTILAGINOMYCOTINA

The class Ustilaginomycetes comprises more than 1400 species of basidiomycetous plant parasites, which are distributed in approximately 70 genera. They occur throughout the world, although many species are restricted to tropical, temperate or arctic regions. Some species of Ustilago and Tilletia, e.g. the barley, wheat or maize smut fungi, are well known because they are of economic importance. Tilletia contraversa is important in the international wheat trade, and 2-5 per cent of the plants in a corn field are generally infected by Ustilago maydis, while up to 80 per cent of a field can be infected if conditions are good for the smut fungus. On the other hand, the galls of U. maydis are considered a delicacy in the Mesoamerican tradition.

Characteristics

- Plant parasitism
- Cellular interaction with primary interactive vesicles
- Cell wall carbohydrate composition with dominance of glucose and absence of xylose
- 5S rRNA secondary structure of type B
- Septal pores without parenthesomes, but in most cases with distinctive tripartite membrane caps or discs
- Life cycle with a parasitic dikaryophase and a saprobic haplophase

Plant Parasitism

In contrast with the Urediniomycetes and Hymenomycetes, the Ustilaginomycetes are ecologically well characterized by their parasitism of vascular plants. Two of the more than 1400 species live on lycophytes, one on ferns, two on conifers, whereas all other Ustilaginomycetes parasitize angiosperms with about 810 species on Poaceae and 170 on Cyperaceae. Interestingly, no species has been reported to parasitize Orchidaceae although this family, with about 20,000 species, is one of the largest groups of the angiosperms. With a few exceptions the teliospore-forming species of the Ustilaginomycetes parasitize nonwoody herbs, whereas those without teliospores prefer woody trees or bushes. However, almost all species sporulate on or in parenchymatic tissues of the hosts. Depending upon the species the sori appear in different organs of the hosts, e.g. in roots, stems, leaves, inflorescences, flowers, anthers, ovaries, seeds etc.

Cellular Interaction

An important apomorphy for the Ustilaginomycetes is the presence of zones of host-parasite interaction with fungal deposits resulting from exocytosis of primary interactive vesicles.

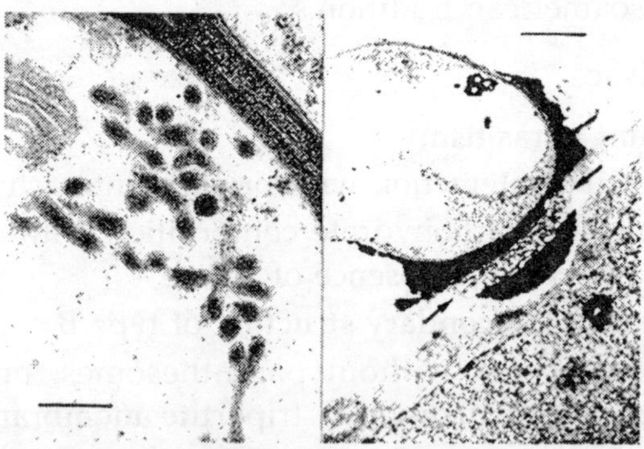

Fig. Transmission electron micrograph showing primary interactive vesicles in Exobasidium pachysporum (Left). Transmission electron micrograph showing a transfer stage between Mycosyrinx cissi (upper cell) and its host (lower cell).

The contents of these vesicles are transferred to the host plasma membrane. Two major types are recognized. (i) Local interaction zones: short-term production of primary interactive vesicles per interaction site results in local interaction zones, and (ii) enlarged interaction zones: continuous production and exocytosis of primary interactive vesicles results in the continuous deposition of fungal material at the whole contact area with the host cell.

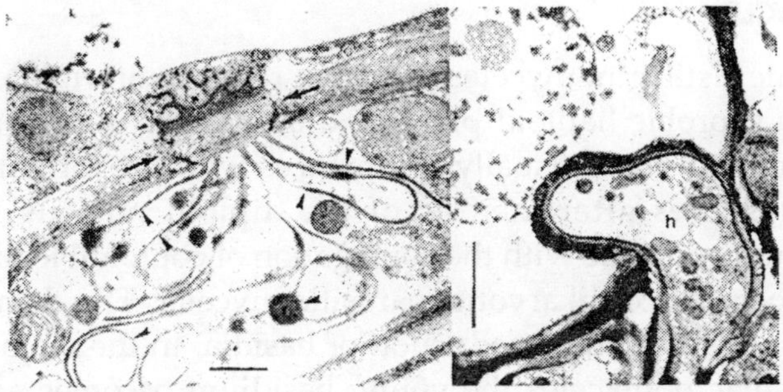

Fig. Transmission electron micrograph showing a local interaction zone (arrows) between Exobasidium pachysporum (lower cell) and its host (upper cell). Note the interaction apparatus (arrowheads) and the deposit at the host cell (left). Transmission electron micrograph showing an enlarged interaction zone between Ustacystis waldsteiniae and its host. The haustorium (h) is encased by electron-opaque material (Right).

Septal Pores

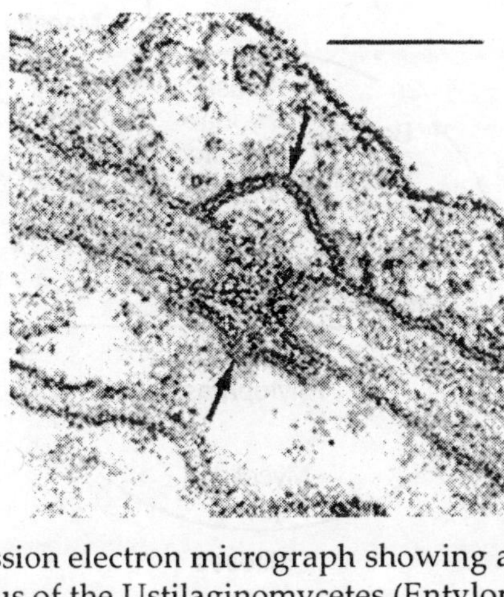

Fig. Transmission electron micrograph showing a typical septal pore apparatus of the Ustilaginomycetes (Entyloma callitrichis) with two membrane caps (arrows).

In contrast with the Hymenomycetes, the septal pores of the Ustilaginomycetes are without multilayered parenthesomes. In contrast with the Urediniomycetes, in most Ustilaginomycetes the septal pores are enclosed by distinctive, tripartite membrane caps or discs.

LIFE CYCLE

The Ustilaginomycetes present a rather uniform life cycle with a saprobic haploid phase and a parasitic dikaryophase. The haploid phase usually commences with the formation of basidiospores after meiosis of the diploid nucleus in the basidium and ends with the conjugation of compatible haploid cells to produce dikaryotic, parasitic mycelia. The dikaryotic phase ends with the production of basidia. In the majority of the Ustilaginomycetes the young basidium becomes a thick-walled teliospore and separates at maturity from the sorus, thus functioning as a dispersal unit. Most of the Ustilaginomycetes are dimorphic, producing a yeast or yeast-like phase in the haploid state. Almost all Ustilaginomycetes multiply mitotically in the saprobic phase, either with yeasts or with ballistoconidia, or with both.

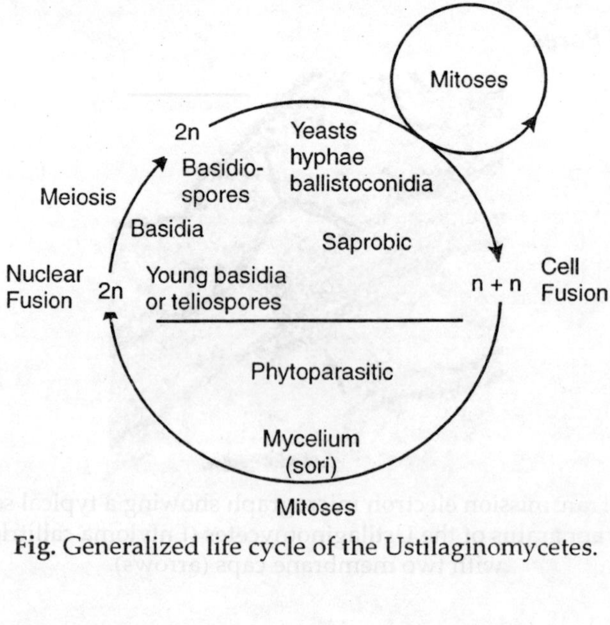

Fig. Generalized life cycle of the Ustilaginomycetes.

The Ustilaginomycetes share most characteristics of the life cycle with the Microbotryales, which traditionally were considered belonging to the Ustilago-group. However, several independent characters show that the microbotryaceous species of the genera Aurantiosporium, Fulvisporium, Liroa, Microbotryum, Sphacelotheca, Ustilentyloma and Zundeliomyces are actually Urediniomycetes.

Phylogenetic Relationships

The smut fungi are traditionally divided into the phragmobasidiate Ustilaginaceae and the holobasidiate Tilletiaceae, which are sometimes treated as separate orders. Difficulties in their classification have been discussed, neither of whom listed higher taxa in this group. Based predominantly on host-parasite interactions and septal pore apparatus, a radical change in the systematics of smut fungi has been proposed. As a result, on the one hand the Microbotryales were excluded from the Ustilaginomycetes and on the other the Exobasidiales s. l., Graphiolales and Microstromatales were included in this group. Cellular interaction and cell wall carbohydrate composition indicate that the class Ustilaginomycetes in this composition is monophyletic, and phylogenetic analyses of rDNA sequences are consistent with this hypothesis.

Ultrastructural and rDNA sequence analyses provide evidence for the existence of three major groups within the Ustilaginomycetes. The basal dichotomy is between the Entorrhizomycetidae and the branch uniting the Ustilaginomycetidae and Exobasidiomycetidae. In contrast with the Ustilaginomycetidae and Exobasidiomycetidae, the septal pores of the Entorrhizomycetidae are not enclosed by tripartite membrane caps. The Ustilaginomycetidae form enlarged interaction zones, whereas the Exobasidiomycetidae form local interaction zones.

Subgroups of Ustilaginomycetes

Lack of membrane bands or caps at the pores and the presence of local interaction zones without interaction apparatus characterize the Entorrhizomycetidae. Entorrhiza is the single genus currently identified of this group.

Presence of enlarged interaction zones characterizes the Ustilaginomycetidae. This statistically well-supported subclass comprises 33 teleomorphic (with a known sexual stage) and one anamorphic (without a known sexual stage) genera, e.g. Anthracoidea living on Cyperaceae, Cintractia living on Cyperaceae and Juncaceae, Doassansiopsis living on mono- and dicots, Farysia living on Cyperaceae, Melanotaenium s. str. living on dicots, Mycosyrinx living on Vitaceae, Pseudozyma (anamorphic genus), Sporisorium living on Poaceae, Thecaphora living on dicots, Urocystis living on mono- and dicots or Ustilago s.str. mainly living on Poaceae.

The Exobasidiomycetidae differ from the Ustilaginomycetidae by forming local interaction zones and from the Entorrhizomycetidae by having membrane caps at the pores. This subclass contains 35 teleomorphic and two anamorphic genera, e.g. Botryoconis living on Lauraceae, Brachybasidium living on Arecaceae, Coniodictyum living on Rhamnaceae, Doassansia living on mono- and dicots, Entyloma living on dicots, Exobasidium living on dicots, Georgefischeria living on Convolvulaceae, Graphiola living on Arecaceae, Malassezia (anamorphic genus), Microstroma living on Juglandaceae and Fagaceae, Tilletia living on Poaceae, Tilletiaria (only known in laboratory) or Tilletiopsis (anamorphic genus).

The Term Smut Fungus

Like the terms agaric, polypore, lichen etc. the term smut fungus circumscribes the organization and life strategy of a fungus, but it is not a taxonomic term. Fungi that look superficially similar to the teliospore-forming members of the Ustilaginomycetes evolved in different fungal groups, e.g. the Microbotryales in the Uredinomycetes or Schroeteria in the Ascomycota.

Pucciniomycotina

Members of the class Urediniomycetes are remarkably diverse, possessing a wide range of structural and ecological characters. One characteristic many of these fungi have in common, however, is their rather miniscule size, making them challenging to locate and study. The apparent lack of unifying structural features of many members of the Urediniomycetes has presented a challenge to understanding their phylogenetic affiliations, and the relationships of many of these disparate taxa have been illuminated since 1995 through the acquisition of molecular sequence, biochemical and subcellular data.

The class contains approximately 7,400 species, distributed among approximately 215 genera, but will eventually be much larger as many cryptic species are recognized. More than 95 per cent of the species and 75 per cent of the genera in this class are placed in the Uredinales, the plant parasitic rust fungi. The next largest orders, Septobasidiales and Microbotryales, collectively constitute approximately 5 per cent of the species and 4 per cent of the genera. Nearly 20 per cent of the rust genera and 60 per cent of the nonrust genera are monotypic (containing only one species). The rust fungi and several of the yeasts have been more extensively studied than have other taxa, and the surprisingly large percentage of monotypic genera may be artificially high due to the limited research on these often obscure fungi.

The rusts are undeniably the most economically important fungi in this clade, as obligate parasites on a wide range of crop plants including cereal grains, legumes, and trees such as coffee, apple, and pine, where they can cause extensive reduction in yield and even host death. Other phytopathogenic Urediniomycetes can be of economic importance, but on a much smaller scale than rust fungi. Species of *Helicobasidium* cause root rot diseases of many economically important plants such as asparagus, beet, mulberry, and pear. *Microbotryum violaceum* causes anther smut of plants in the Caryophyllaceae, while smut on buckwheat is caused by *Sphacelotheca fagopyri*. These smuts are not closely related to the smuts in the Ustilaginomycetes with which they were confused until molecular sequence data became available.

CLASSES OF BASIDIOMYCOTA

The Basidiomycota had traditionally been divided into 2 obsolete classes, the Homobasidiomycetes (including true mushrooms); and the Heterobasidiomycetes (the Jelly, Rust and Smut fungi). Previously the entire Basidiomycota were called Basidiomycetes, an invalid class level name coined in 1959 as a counterpart to the Ascomycetes, when neither of these taxa were recognized as phyla. The terms basidiomycetes and ascomycetes are frequently used loosely to refer to Basidiomycota and Ascomycota. They are often abbreviated to "basidios" and "ascos" as mycological slang.

Rusts

Rusts (Pucciniales, previously known as Uredinales) at their greatest complexity produce five different types of spores on two different hosts in two unrelated host families. Such rusts are heteroecious (requiring 2 hosts) and macrocyclic (producing all 5 spores types). By convention the stages and spore states are numbered by Roman numerals. Typically, basidiospores infect host one, the mycelium forms pycnidia, called spermagonia, which are miniature, flask-shaped, hollow, submicroscopic bodies embedded in host tissue (such as a leaf). This stage, numbered "0", produces single-celled, minute spores that ooze out in a sweet liquid and that act as nonmotile spermatia, and also protruding receptive hyphae. Insects and probably other vectors such as rain carry the spermatia from spermagonia to spermagonia, cross inoculating the mating types. Neither thallus is male or female. Once crossed, the dikaryons are established and a second spore stage is formed, numbered "I" and called aecia, which form dikaryotic aeciospores in dry chains in inverted cup-shaped bodies embedded in host tissue. These aeciospores then infect the second host genus and cannot infect the host on which they are formed (in macrocyclic rusts). On the second host a repeating spore stage is formed, numbered "II", the uredospores in dry pustules called uredinia. Urediospores are dikaryotic and can infect the same host that produced them. They repeatedly infect this host over the growing season. At the end of the season, a fourth spore type, the teliospore, is formed. It is thicker-walled and serves to overwinter or to survive other harsh conditions. It does not continue the infection process, rather it remains dormant for a period and then germinates to form basidia (stage "IV"), sometimes called a promycelium. In the Pucciniales, the basidia are cylindrical and become 3-septate after meiosis, with each of the 4 cells bearing one basidiospore each. The basidospores disperse and start the infection process on host 1 again. Autoecious rusts complete their life-cycles on one host intead of two, and microcyclic rusts cut out one or more stages.

Smuts

The characteristic part of the life-cycle of smuts is the thick-walled, often darkly pigmented, ornate, teliospore that serves to survive harsh conditions such as overwintering and also serves to help disperse the fungus as dry diaspores. The teliospores are initially dikaryotic but become diploid via karyogamy. Meiosis takes place at the time of germination. A promycelim is formed that consists to a short hypha (equated to a basidium). In some smuts such as *Ustilago maydis* the nuclei migrate into the promycelium that becomes septate, and haploid yeast-like conidia/basidiospores sometimes called sporidia, bud off laterally from each cell. In various smuts, the yeast phase may proliferate, or they may fuse, or they may infect plant tissue and become hyphal. In other smuts, such as *Tilletia caries*, the elongated haploid basidiospores form apically, often in compatible pairs that fuse centrally resulting in "H"-shaped diaspores which are by then dikaryotic. Dikaryotic conidia may then form. Eventually the host is infected by infectious hyphae. Teliospores form in host tissue. Many variations on these general themes occur.

Dimorphic Basidiomycota are smuts with both a yeast phase and an infectious hyphal state are examples of dimorphic Basidiomycota. In plant parasitic taxa, the saprophytic phase is normally the yeast while the infectious stage is hyphal. However, there are examples of animal and human parasites where the species are dimorphic but it is the yeast-like state that is infectious. The genus *Filobasidiella* forms basidia on hyphae but the main infectious stage is more commonly known by the anamorphic yeast name *Cryptococcus*, e.g. *Cryptococcus neoformans* and *Cryptococcus gattii*.

The dimorphic Basidiomycota with yeast stages and the pleiomorphic rusts are examples of fungi with anamorphs, which are the asexual stages. Some Basidiomycota are only known as anamorphs. Many are yeasts, collectively called basidiomycetous yeasts to differentiate them from ascomycetous yeasts in the Ascomycota. Aside from yeast anamorphs, and uredinia, aecia and pycnidia, some Basidiomycota form other distinctive anamorphs as parts of their life-cycles. Examples are *Collybia tuberosa* with its apple-seed-shaped and coloured sclerotium, *Dendrocollybia racemosa* with its sclerotium and its *Tilachlidiopsis racemosa* conidia, *Armillaria* with their rhizomorphs, *Hohenbuehelia* with their *Nematoctonus* nematode infectious, state and the coffee leaf parasite, *Mycena citricolour* and its *Decapitatus flavidus* propagules called gemmae.

ANTI-CANCER EFFECT OF BASIDIOMYCETES MUSHROOMS

The global awareness of cancer as the second largestcause of death in people of various ages and racial background has lead to so much research efforts and clinical studies in the fight against the disease. The prevention of cancer by the ingestion of chemical agents aimed at minimizing the risk of carcinogenesis has greatly reduced the morbidity and mortality rates. These agents include non-steroid anti-inflammatory drugs (NSAID) such as aspirin, sulindac, piroxican and indomethacin, which inhibit cyclooxygenase (COX).

Cyclooxygenase inhibition is necessary because it catalyzes the conversion of arachidonic acid to pro-inflammatory substances such as prostaglandin, which can stimulate growth of tumor cells and suppress immune surveillance. It also activates carcinogens to take up forms that damage the genetic material.

The curious question, which everyone asks, is " what really causes cancer." The immediate response from most people will be tobacco smoking, alcohol consumption, excess use of caffeine and other drugs, sunshine, infection from such oncogenic virus like cervical papillomaviruses, adenoviruses Karposis sarcoma (HSV) or exposure to asbestos.

These obviously are implicated as causal agents of mammalian cancers. However a large population of people are often exposed to these agents with only very few being affected which tell us that these may not really be the root cause of cancer.

Gibbs explained that the immediate cause of cancer must be some combination of insults and accidents that induce normal cells in a healthy human body to turn malignant, growing uncontrollably in unnatural places. The elucidation of DNA as the genetic material ushered in the era of molecular genetics and cancer is now seen as a result of cumulative mutations that alter specific locations in a cell's DNA and thus change the particular protein encoded by cancer-related genes at those locations. These mutations affect two kinds of cancer genes namely: tumor suppressors and oncogenes. Tumor suppressors normally restrain the cell's ability to divide and mutations on this gene render it inactive. Oncogenes on the other hand stimulate cell growth and cell division. Mutations on the oncogenes make them permanently active thus promoting cell growth and progressive changes leading to the increase of the cancer genes. Consequently cancer cells continue to divide even in situations in which normal cells will usually wait for a special chemical transduction signal. The tumor cells would ignore such stop signals that are sent out by adjacent tissues. Cancer cells also have the character of immortality even in vitro whereas normal cells stop dividing after 50-70 generations and undergoes a programmed cell death (Apoptosis). Cancer cells continue to grow invading nearby tissues and metastasizing to distant parts of the body. Metastasis is the most lethal aspect of carcinogenesis. The phenomena is such that before a tumor is discovered and removed by surgery a few cells of the tumor would detach from the initial mass, float through the circulation system and start a new colony in a different organ from the one that gave birth to it.

The theory that cancer results from cumulative mutations that alter specific genes in the DNA has come under great challenge recently. Biologists now trace the cause of tumor to include other abnormalities at work inside the cell nucleus that though not yet malignant is prone to become so. Recent hypotheses advanced to challenge the earlier one include the hypothesis that a breakdown in DNA duplication or repair leads to many thousands of random mutations in cells. Secondly that damage to a few "master" genes mangles the chromosomes, which then becomes dangerous, plunging the cells into a ploidal confusion and chaos.

Finally that the abnormal number of chromosomes in a cell may be the milestone on the road to cancer.

However, there is great expectation that one day science will produce a definitive answer to the root cause of cancer. It may probably be a very complicated answer or a simple one, which may force us to shift our hope from curative drugs to chemopreventives.

It has been difficult to develop specific remedies against cancer as with the development of vaccine against viral infections and antibiotics. Cancer cells originate from normal cells and a novel drug able to select and destroy only affected cells preventing carcinogenesis without injury to normal cell would be an ideal chemotherapy against cancer. This is a very difficult task facing cancer researchers. Recent investigations has been channeled on the development of immunotherapy to target and remove cancer cells as well as on substances such as immunopotentiators, immunoinitiators and biological response modulators (BRM) that act to prevent carcinogenesis and induce carcinostasis.

Mushrooms belong to this group of immunoceuticals by their mode of action. The use of medicinal mushrooms in the fight against cancer is known for a very long time in Korea, China, Japan, Russia, USA and Canada. An old Japanese legend reports that wild monkeys rarely experience cancer, high blood pressure, or diabetes. The legend suggests that perhaps it is due to some extent their consumption of wild mushrooms. This legend in Japan may have help spur research into the role of mushroom in medicine. These mushrooms belonging to the family Polyporaceae have been effective against esophageal, stomach, prostate and lung cancers. Lucas demonstrated the anti-tumor effect of higher basidiomyces (specifically extracts of fruiting bodies of Boletus edulis). Yohida isolated from Lampteromyces japonicus (kowamura) Sing, an agent active against Ehrlich carcinoma of the mouse. Gregory experimented on more than 7000 cultures of higher basidiomyces for anti-tumor activity against rodent tumor systems. Positive inhibitory effects were obtained using fermentation media materials against sarcoma 180, mammary adenocarcinoma 755, and leukemia L-1210.

Ikegawa reported that an essence obtained from the fruit body of edible mushrooms exhibited remarkable host-mediatory anti-tumor activity against grafted cancer in animals such as sarcoma 180. Daba and Daba et al. reported that Pleurotus ostreatus mushrooms cultivated on date waste posses a potent antitumor activity against Ehrlich ascites carcinoma. Further biochemical studies were carried out by the authors on the effect of mushrooms and isolated polysaccharides on the tumors transplanted in mice. The anti-tumor essence was later discovered to be a type of β-D-glucan, a polysaccharide yielding D-glucose only by acid hydrolysis. β-D-glucan has a different mode of action from the conventional chemotherapeutic agents in that it is immunotherapeutic. Inhibiting the growth of cancer cells by activating and reinforcing the host immune system. This present review elucidates the anti-tumor polysaccharides from mushrooms, their chemical structures, results obtained from experimental and clinical trails, and their mode of action.

COMPOSITION OF MUSHROOM ANTITUMOR POLYSACCHARIDES

Polysaccharides are polymers of sugars (monosaccharides) joined to each other by glycosidic linkages. These are very complex molecules because sometimes covalent bonds occur between many pairs of carbon atoms. Consequently one sugar unit can be joined to more than two other sugars, which results in the formation of highly branched enormous macromolecules. Polysaccharides are a structurally diverse class of macromolecules able to offer the highest capacity for carrying biological information due to a high potential for structural variability.

Whereas the nucleotides and amino acids in nucleic acids and proteins effectively, interconnect in only one way, the monosaccharide units in polysaccharides can interconnect at several points to form a wide variety of branched or linear structures. This high potential for structural variability in polysaccharides gives the necessary flexibility to the precise regulatory mechanisms of various cell-cell interactions in higher organisms. The polysaccharides of mushrooms occur mostly as glucans. Some of which are linked by β-(1-3), (1-6) glycosidic bonds and α-(1-3) glycosidic bonds but many are true heteroglycans. Most often there is a main chain, which is either β(1-3), β(1-4) or mixed β(1-3), β(1-4) with β(1-6) side chains. Hetero- β-D-glucans, which are linear polymers of glucose with other D-monosaccharides, can have anticancer activity but α-D-glucans from mushroom usually lack anticancer activity. Heteroglucan side chains contain glucuronic acid, galactose, mannose, arabinose or xylose as a main component or in different combinations. Glycans are polysaccharides containing units other than glucose in their backbone. A wide range of antitumor or immunostimulating polysaccharides of different chemical structure from higher Basidiomyces mushrooms has been investigated. Some correlation has been drawn between the chemical structure and antitumor activities of mushroom polysaccharides. A wide range of glycans extending from homopolymers to highly complex heteropolymers exhibits antitumor activity. Differences in activity can all be correlated with ability of the polysaccharide molecule to solubilize in water, size of the molecules, branching rate and form. Such structural features as β-(1-3) linkages in the backbone (main chain) of the glucan and additional β-(1-6)-branch points are needed for antitumor activity. β-glucans with only (1-6) glycosidic linkages have little or no activity. Higher molecular weight glucans have been reported by Mizuno et al. (1996) and Mizuno (1999) to be more effective than those of low molecular weight against tumors.

There is a broad similarity in the various methods that have been developed for extraction of anti-cancer polysaccharides from mushrooms. Usually dried mushroom in powdered form is heated in 80 per cent ethanol to eliminate low molecular weight substances. Crude fractions are obtained from the remaining ethanol extracts by further extraction with water, 1 per cent ammonium oxalate and 5 per cent sodium hydroxide. The polysaccharides are then fractionally purified by a combination of techniques, including ethanol concentration, fractional precipitation, ion-exchange chromatography, gel filtration and affinity chromatography. A recent study by Yap and Ng (2001) has established a more efficient procedure for the extraction of β-glucan from mushroom. The β-glucan is isolated through ethanol precipitation and freeze-dried in liquid nitrogen. Purity testing using a carbohydrate analysis column produced 87.65 per cent purity.

MUSHROOMS POLYSACCHARIDES IN THE TREATMENT OF CANCER

Immunoceuticals are substances having immunotherapeutic efficacy when administered orally. More than 50 mushrooms species have yielded potential immunoceuticals that exhibit anticancer activity invitro or in animal models. Six of these polysaccharides that have been investigated in human cancers include Lentinan, Schizophyllan, Active hexose correlated compounds (AHCC), Maitake D-fraction, polysaccharide-K and Polysaccharide-P.

LENTINAN

Lentinan, produced from Shiitake mushroom, Lentinus edodes, is a β(1-3), β (1-6) glucan. There is an immense literature related to the anticancer effect of lentinan on animals and human carcinomas. It was first isolated and studied by Chihara who demonstrated that its antitumor effects were greater than other mushroom polysaccharides. Maeda et al. however reported that lentinan was active for some but not all types of tumors. There have been numerous clinical trials of lentinan in Japan, and the drug is now manufactured and sold by several pharmaceutical companies.

Lentinan has proved successful in prolonging the overall survival of cancer patients, especially those with gastric and colourectal carcinomas. In patients with recurrent gastric cancer, tumor responses and prolonged median survival were also noted. In a randomized controlled study of patients treated with tegafur or a combination of lentinan and tegafur overall survival was significantly prolonged in the lentinan plus tegafur group. Overall more patients with the combined therapy appeared to survive longer: 19.5 per cent survived more than one year 10.4 per cent more than two years and 6.5 per cent more than three years. Using the criteria of the Japan society for cancer therapy for evaluation of clinical effects of cancer chemotherapy on solid tumours, patients treated with lentinan had a significantly higher response rate (14.9 per cent than patients in the control arm. Few adverse reactions to lentinan have been noted. In a detailed study of 469 patients, 32 (6.8 per cent) experienced an adverse reaction, only 2 patients required discontinuation of treatment due to unacceptable tolerance. Perhaps the most intriguing aspects of lentinan use in conjunction with other chemotherapeutic agents is its apparent ability to greatly reduce the debilitating effects of the chemotherapy, e.g. nausea, pain, hair loss and lowered immune status.

SCHIZOPHYLLAN

This polysaccharide derived from the mushroom Schizophyllum commune has been shown to be cytostatic in sarcoma 180 tumor xenographs. The survival of these xenographs was not affected by pretreatment with schizophyllan. Schizophyllan had no effect on the survival of sarcoma 37, Ehrlich carcinoma or Yoshida sarcoma ascites tumors.

Various clinical trials have been carried out in Japan, although many are not blinded. Despite this schizophyllan has been approved for clinical use in Japan. Early clinical studies with schizophyllan in combination with conventional chemotherapy (tegafur or mitomycin C and 5-fluorouracil) in a randomized controlled study of 367 patients with recurrent and inoperable gastric cancer resulted in a significant increase in median survival. However, a similar study was unable to confirm this apparent success with schizophyllan. Recently schizophyllan has also been shown to increase overall survival of patients with head and neck cancers. In a randomized controlled study of schizophyllan in combination with radiotherapy, schizophyllan significantly prolonged the overall survival of stage II cervical cancer patients but not stage III. In a prospective, randomized clinical trial involving 312 patients treated with surgery, radiotherapy, chemotherapy (fluorouracil) and schizophyllan in various combinations, patients treated with schizophyllan had a better overall survival than patients who had not received the polysaccharides. Schizophyllan is currently produced commercially by several Japanese pharmaceutical companies.

ACTIVE HEXOSE CORRELATED COMPOUND (AHCC)

AHCC is a proprietary extract prepared from co-cultured mycelia of several species of basidiomycete mushrooms, including Shiitake (Lentinus edodes). Animal research and preliminary human studies indicated AHCC has anticancer efficacy. Beginning in 1992, Kamiyama conducted a trial in Japan to evaluate the preventive effect of AHCC against recurrence of hepatocellular carcinoma following surgical resection. The investigators reported that after one year the AHCC group showed a significant higher survival rate than the control group, as well as significant lowering of certain tumor markers in the serum. The AHCC Research Association was formed in Japan in 1996 to foster the development of AHCC as an anticancer therapy. In their circulating abstracts they report that of 300 cancer patients administered AHCC, 58 were effectively treated, with 46 showing complete or partial regression and 12 experiencing no change of tumor size.

Maitake Extracts

Several studies have shown that B-D-glucan derived from Grifola frondosa (also known as Maitake) have strong antitumor activity in xenographs and there have also been limited number of clinical trials. More recently, a highly purified extract, β-glucan (B-1,6 glucan branched with a β 1-3-linkage) (Grifron-D, GD) has become available. GD has considerable immunomodulating and antitumor activities in animal models, and is orally bioavailable. Maitake D-fraction and crude Maitake powder have demonstrated remarkable inhibition of metastasis in a mouse model, especially in the prevention of hepatic metastases, which in one series of experiments was reduced by 81 per cent (Maitake powder) to 91 per cent (D-fraction). GD has been shown to have a cytotoxic affect on human prostate cancer cells (PC9) in vitro, possibly acting through oxidative stress, and causing 95 per cent cell death by an apoptosis. Vitamin C addition reduced the effective level of GD required.

Jones reported an early pilot study from China involving 63 cancer patients which showed a response rate (partial and complete) against solid tumour at 95 per cent and for leukaemia 90 per cent. Nanba observed tumour regression or significant symptomatic improvement in 11 out of 15 hepatocellular carcinomas with D-fraction plus Maitake. When D-fraction plus Maitake was combined with chemotherapy, the overall response rates were increased by 12-28 per cent when results from cancer types were combined. The Food and Drug Administrarion has approved Grifon-D (GD) fòr trial under an Investigational New Drug Application (IND) for patients with advanced cancer are currently underway at various Institutions.

Development of Carcinostatic Drugs

Three polysaccharide based carcinostatic (immunotherap eutic) agents, Krestin, Lentinan and Sonifilan, have already been developed from mushroom. These are used currently in the treatment of cancer of the digestive organs, lung and breast, as well as cancer of the stomach and cervical cancer respectively. Several mushroom species belonging to the polyporaceae family are now regarded as the next drug producers. Mushroom polysaccharides are also expected to be developed into multipurpose medicines that are not only carcinostatic but also anti-inflammatory, antiviral against AIDS), hypoglycaemic and antithrombotic.

MECHANISM OF ACTION OF MUSHROOM POLYSACCHARIDES

The Egyptians as far back as 3000 BC believed that mushroom were a sacred food that prolonged life. A mummified 5000-year old "Ice-man" found in the mountains of Europe carried a medicine kit of dried mushrooms. Indeed the oldest written record of mushrooms as medicines is in Indian medical treatise from 3000BC. Of significant relevance and importance is the ability of particular mushrooms-derived compounds to modulate the human immune response and to inhibit certain tumour growths. Medicinal mushroom research has focused on discovery of compounds that can modulate positively or negatively the biologic response of immune cells. Those compounds, which appear to stimulate the human immune response, are being sought for the treatment of cancer, immunodeficiency disease or for generalised immuno-suppression following drug treatment. They are also sought for combination therapy with antibiotics and as adjuncts for vaccines. Wasser reported that mushroom polysaccharides are regarded as biological response modifiers (BRM). This basically means that they cause no harm and place no additional stress on the body, but help the body to adapt to various environmental and biological stresses. Mushroom polysaccharides support some or all of the major systems of the body, including nervous, hormonal and immune systems as well as regulatory functions. The polysaccharides from mushrooms do not attack cancer cells directly, but produce their antitumour effects by activating different immune response in the host.

Hobbs reported that L. edodes produces two bioactive preparations, which are efficient immune modulators, mycelium extract and Lentinan. These two bioactive polymers appear to act as host defence potentiators restoring and enhancing the responsiveness of host cells to lymphocytokines, hormone and other biologically active substances. The immunopotentiation has been shown to occur by stimulating the maturation, differentiation or proliferation of cells involved in host defence mechanism. Chihara et al. reported that Lentinan increase host's resistance against various kinds of cancer and has the potential to restore the immune function of affected subjects.

Many of these pathways stimulated by Lentinan are illustrated by Chihara et al.. The interaction of bLentinan with many kinds of immune cells was not known until recently. Ross et al. provided an insight into receptor binding in immune cells by β-glucan from fungi and further showed that β-glucan from yeast bind to iC3b- receptors (CR3, CD11b/CD18) of phagocytic and natural killer (NK) cells. When this happens, it will stimulate phagocytosis and/or cytotoxic degranulation. Lentinan has also been shown to stimulate peripheral blood lymphocytes in vitro to increase interleukin-2-mediated LAK cell (lymphokine-activated killer cell) and NK cell activity at levels achievable in vivo by administration of clinical doses of Lentinan. This observation was made using the blood of healthy donors and cancer patients. Lentinan has also been shown to inhibit suppressor T cells activity in vivo and to increase the ratio of activated T cells and cytotoxic T cells in the spleen when administered to gastric cancer patients undergoing chemotherapy. Many interesting biological activities of Lentinan including increase in the activation of non- specific inflammatory response such as acute phase protein production; vascular dilation and haemorrhage-inducing factor in vivo, activation and generation of helper and cytotoxic T cells.

CONCLUSION

Mushroom polysaccharides offer a lot of hope for cancer patients and sufferers of many devastating diseases. A fundamental principle in oriental medicine is to regulate homeostasis of the whole body and to bring the disease person to his or her normal state.

Potentiation of the physiological constitution in favour of host defence results in the activation of many vitally important cells for the maintenance of homeostasis. A variety of polysaccharides from a number of mushroom varieties have been demonstrated to enhance the immune system. All of these have shown significant antitumour activity as a result of their ability to activate the host immune system rather than direct cytotoxicity. The mushroom polysaccharides appear to be well tolerated and compatible with chemotherapy and radiation therapy. However, studies that identifies the molecular mechanisms that occur in specific immune modulation by mushroom polysaccharides such as receptors and what downstream events are triggered by the binding of these polymers to their target cells are urgently needed.

4

Mycoplasma

Mycoplasma is a genus of bacteria that lack a cell wall. Because they lack a cell wall, they are unaffected by some antibiotics such as penicillin or other beta-lactam antibiotics that target cell wall synthesis. They can be parasitic or saprophytic. Several species are pathogenic in humans, including *M. pneumoniae*, which is an important cause of pneumonia and other respiratory disorders, and *M. genitalium*, which is believed to be involved in pelvic inflammatory diseases. They may cause or contribute to some cancers.

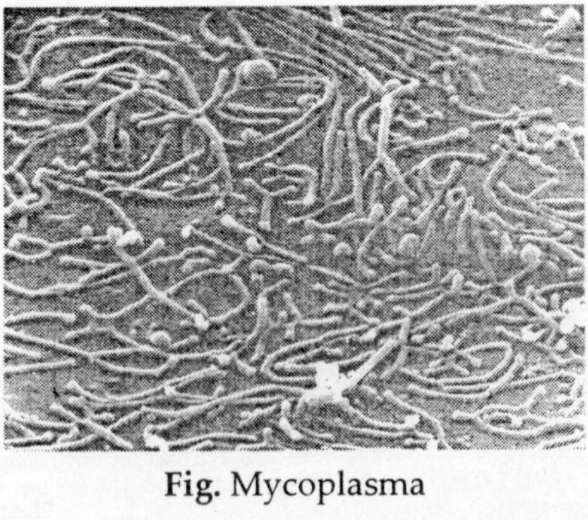

Fig. Mycoplasma

The genus *Mycoplasma* is one of several genera within the class *Mollicutes*. Mollicutes are bacteria which have small genomes, lack a cell wall and have low a GC-content (18-40 molpercent). There are over 100 recognized species of the genus *Mycoplasma*. Their genome size ranges from 0.58 - 1.38 megabase-pairs. Mollicutes are parasites or commensals of humans, animals (including insects), and plants; the genus *Mycoplasma* is by definition restricted to vertebrate hosts. Cholesterol is required for the growth of species of the genus *Mycoplasma* as well as certain other genera of mollicutes. Their optimum growth temperature is often the temperature of their host if warmbodied (e.g. 37 degrees Celsius in humans) or ambient temperature if the host is unable to regulate its own internal temperature. Analysis of 16S ribosomal RNA sequences as well as gene content strongly suggest that the mollicutes, including the mycoplasmas, are closely related to either the *Lactobacillus* or the *Clostridium* branch of the phylogenetic tree (Firmicutes *sensu stricto*).

Mycoplasmas are often found in research laboratories as contaminants in cell culture. Mycoplasmal cell culture contamination occurs due to careless working or contaminated cell culture medium ingredients. The Mycoplasma cell is usually smaller than 1 μm and they are therefore difficult to detect with a conventional microscope. Their presence in cell cultures or media can skew experimental results.

MYCOPLASMA HISTORY AND GENERAL CHARACTERISTICS

The bacteria of the genus *Mycoplasma* (trivial name: mycoplasmas) and their close relatives are largely characterised by lack of a cell wall. In spite of this deficiency, the shapes of these cells often conform to one of several possibilities with varying degrees of intricacy. The members of the genus *Spiroplasma* assume an elongated helical shape without the aid of a rigid structural cell envelope. These cell shapes presumably contribute to the ability of mycoplasmas to thrive in their respective environments. *M. pneumoniae* cells possess an extended 'arm' protruding from a coccoid cell body, which is involved in the attachment of this pathogenic bacterium to the tissue of its human host, in movement along solid surfaces, and in cell division. *M. pneumoniae* cells are of small size and somewhat pleomorphic, but with a rough shape in longitudinal cross-section resembling that of a round-bottomed flask.

Mycoplasmas are extremely unusual among bacteria in that most require sterols for the stability of their cytoplasmic membrane. Sterols are acquired from the environment, usually as cholesterol from the animal host. Mycoplasmas also generally possess a relatively small genome of 0.6-1.35 megabases reflecting their drastically reduced biosynthetic capabilities and parasitic lifestyle, with a low mol percent G+C ranging from 18-40 per cent. This is coupled with the use of an alternate genetic code where the codon UGA is preferred to encode the amino acid tryptophan instead of the usual opal stop.

In 1898 Nocard and Roux reported the cultivation of the causative agent of contagious bovine pleuropneumonia (CBPP), at the time a grave disease in agriculture and today a concern of cattle ranchers particularly in Africa and Southern Europe, and of customs officials elsewhere. The disease is caused by *M. mycoides* subsp. mycoides SC (small-colony type), and the work of Nocard and Roux represented the first isolation of a mycoplasma. Its culture was difficult because of the complex growth requirements. These researchers succeeded by inoculating a semi-permeable pouch of sterile medium with pulmonary fluid from an infected animal and depositing this pouch intraperitoneally into a live rabbit. After fifteen to twenty days, the recovered pouch had an opacity that an uninoculated control lacked. This turbid broth could then be used to inoculate a second and third round and subsequently introduced into a healthy animal, causing disease. However this did not work if the material was heated, indicating a biological agent at work. Uninoculated media in the pouch, after removal from the rabbit, could be used to grow the organism in vitro, demonstrating cell-free culture and ruling out viral causes, although this was not fully appreciated at the time. The name *Mycoplasma,* from the Greek mykes (fungus) and plasma (formed), was proposed in the 1950's, replacing the term pleuropneumonia-like organisms (PPLO) referring to organisms similar to the causative agent of CBPP. It was later found that the fungus-like growth pattern of *M. mycodies* is unique to that species.

This confusion about mycoplasmas and virus would surface again 50 years later when Eaton and colleagues cultured the causative agent of human primary atypical pneumonia (PAP) or 'walking pneumonia.' This agent could be grown in chicken embryos and passed through a filter that excluded normal bacteria, but could not be observed by the high magnification light microscopy of the day, and caused disease that could not be treated with the popular antimicrobials sulphonamides and penicillin. Eaton did consider the possibility the disease was caused by a mycoplasma, but the agent did not grow on the standard PPLO media of the time. These observations led to the conclusion that PAP had a viral etiology. Research at that time showed the cultured agent could induce disease in experimentally infected cotton rats and hamsters. In spite of controversy at the time about whether the researchers had truly isolated the causative agent of PAP (based largely on the unusual immunological response of patients with PAP), in retrospect their evidence along with that of colleagues and competitors appears to have been quite conclusive. In the early 1960's, there were reports linking Eaton's Agent to the PPLOs or mycoplasmas, well known then as parasites of cattle and rodents, using sensitivity to antimicrobial compounds (i.e. organic gold salt). The ability to grow Eaton's Agent, now known as *Mycoplasma pneumoniae,* in cell free media allowed an explosion of research into what had overnight become the most medically important mycoplasma (from a human-centric perspective) and what was to become the most studied mycoplasma.

Recent advances in molecular biology and genomics have brought the genetically simple mycoplasmas, particularly *M. pneumoniae* and its close relative *M. genitalium,* to a larger audience. The second published complete bacterial genome sequence was that of *M. genitalium,* which has the smallest genome of any free-living organism. The *M. pneumoniae* genome sequence was published soon after and was the first genome sequence determined by primer walking of a cosmid library instead of the whole-genome shotgun method. Mycoplasma genomics and proteomics continue in efforts to understand the so-called minimal cell, catalog the entire protein content of a cell, and generally continue to take advantage of the small genome of these organisms to understand broad biological concepts.

MYCOPLASMA TAXONOMY AND PHYLOGENY

The medical and agricultural importance of members of the genus *Mycoplasma* and related genera has led to the extensive cataloging of many of these organisms by culture, serology, and small subunit rRNA gene and whole genome sequencing. A recent focus in the sub-discipline of molecular phylogenetics has both clarified and confused certain aspects of the organization of the class *Mollicutes*, and while a truce of sorts has been reached, the area is still somewhat of a moving target.

The name mollicutes is derived from the Latin mollis (soft) and cutes (skin), and all of these bacteria do lack a cell wall and the genetic capability to synthesize peptidoglycan. While the trivial name 'mycoplasmas' has commonly denoted all members of this class, this usage is somewhat imprecise and will not be used as such here. Despite the lack of a cell wall, Mycoplasma and relatives have been classified into the phylum Firmicutes consisting of low G+C Gram-positive bacteria such as Clostridium, Lactobacillus, and Streptococcus based on 16S rRNA gene analysis. The cultured members of Mollicutes are currently arranged into four orders: Acholeplasmatales, Anaeroplasmatales, Entomoplasmatales, and Mycoplasmatales. The order Mycoplasmatales contains a single family, Mycoplasmataceae, which contains two genera: Mycoplasma and Ureaplasma. Historically, the description of a bacterium lacking a cell wall was sufficient to classify it to the genus *Mycoplasma* and as such it is the oldest and largest genus of the class with about half of the class' species (107 validly described) each usually limited to a specific host and with many hosts harboring more than one species, some pathogenic and some commensal. In later studies, many of these species were found to be phylogenetically distributed among at least three separate orders. In fact, the type species, *M. mycoides* would rightly be classified with the genus *Spiroplasma* in the order *Entomoplasmatales*. This and other discrepancies will likely remain unresolved because of the extreme confusion that change could engender among the medical and agricultural communities. The bulk of the species in the genus *Mycoplasma* are divided into two non-taxonomic groups based on 16S rRNA gene sequences, hominis and pneumoniae. The hominis group contains the phylogenetic clusters of *M. bovis, M. pulmonis,* and *M. hominis* among others. The pneumoniae group contains the clusters of *M. muris, M. fastidiosum, U. urealyticum,* the uncultured Haemotrophic mollicutes, haemoplasmas (formally *Haemobartonella*), and the *M. pneumoniae* cluster. This cluster contains the species (and the usual or likely host) *M. alvi* (bovine), *M. gallisepticum* (avian), *M. genitalium* (human), *M. imitans* (avian), *M. pirum* (uncertain/human), *M. testudinis* (tortoises), and *M. pneumoniae* (human). Most if not all of these species share some otherwise unique characteristics including an attachment organelle, homologs of the *M. pneumoniae* cytadherence-accessory proteins, and specialized modifications of the cell-division apparatus.

A detailed analysis of the 16S rRNA genes from the order *Mollicutes* by Maniloff has given rise to a view of the evolution of these bacteria that includes an estimate of the time-scale for the emergence of some groups or features. This analysis suggests that about 600 million years ago (MYA), late in the Proterozoic era, Mollicutes branched away from the low G+C Gram-positive ancestor of the streptococci, losing their cell wall. At this time on Earth, molecular oxygen was present in the atmosphere at 1 per cent, and the fossil record shows that multicellular marine animals had recently spread in the Cambrian explosion. One hundred million years later the requirement for sterols in the cytoplasmic membrane evolved along with the change to the alternate genetic code. Also, the ancestor of the the genera Spiroplasma and Entomoplasma (primarily plant and insect pathogens) and Mycoplasma emerged at this time and would itself diverge into the Spiroplasma-Entomoplasma and Mycoplasma lineages approximately 100 million years after that. This diversity coincided with the origin of land plants 500 MYA. It appears that the calculated rate of evolution for the Mycoplasma group increased several fold about 190 MYA, soon after the appearance of vertebrates, while the *Spiroplasama-Entomoplasma* ancestor continued to evolve at the previously shared slower rate until about 100 MYA, when angiosperms and their associated pollinating insects appeared. Then the evolution rate of these bacteria appears to have also increased significantly. This is an attractive hypothesis, but while it tracks the emergence of several of the unusual characteristics of *Mycoplasma* and related organisms, it does not address the selective pressures driving their evolution, except perhaps the widespread close association of a parasite with a specific host. The advantages of a reduced genome, cell wall-less structure, and alternate genetic code remain murky.

METABOLISM

The parasitic mycoplasmas have truncated respiratory systems, lacking quinones and cytochromes. Another indication for the simplicity of the electron transport chain is the finding that the reduced nicotinamide adenine dinucleotide (NADH) oxidase activity is cytoplasmic. Complex electron transport chains are usually membrane bound, since they depend on the spatial organization of their components. Ruling out oxidative phosphorylation as an ATP-generating system leaves only two proven ways of ATP generation, both based on substrate level phosphorylation. The major source for ATP is the arginine dihydrolase pathway.

$$\text{Arginine} + H_2O \xrightarrow{\text{arginine deaminase}} \text{citrulline} + NH_3$$

$$\text{Citrulline + inorganic orthophosphate} \xrightarrow{\text{ornithine carbamoyltransferase}} \text{ornithine+carbamoylPO}$$

$$\text{Carbamoyl } PO_4 + \text{ADP} \xrightarrow{\textit{carbamate kinase}} \text{ATP} + CO_2 + NH_3$$

Another mechanism for ATP generation is:

$$\text{Acetyl CoA + inorganic orthophosphate} \xrightarrow{\textit{phosphate acetyltransferase}} \text{acetyl } PO_4 + \text{CoA}$$

$$\text{Acetyl } PO_4 + \text{ADP} \xrightarrow{\textit{acetate kinase}} \text{Acetate + ATP}$$

Acetyl CoA is produced by oxidative decarboxylation of pyruvate.

A few species derive their energy from the degradation of glucose or the hydrolysis of urea. All species synthesize DNA, RNA, lipids and proteins. However, it is not known if they can synthesize amino acids. Those species that require sterols incorporate these sterols (mainly cholesterol) into the cell membrane up to concentrations of 65 per cent.

DISEASES

The human diseases caused by mycoplasmas.

Disease or Symptom	Agent	Host
Primary atypical pneumonia	Mycoplasma pneumoniae	Man
Non-gonococcal urethritis (NGU)	Mycoplasma genitalium	Man
NGU	Ureaplasma urealyticum	Man
Stillbirth	Mycoplasma hominis	Man
Spontaneous abortion	Mycoplasma hominis	Man
Infertility	Mycoplasma hominis	Man

TREATMENT OF MYCOPLASMA

Mycoplasmas are sensitive to various antibiotics like the tetracyclines, macrolides, and the quinolones, but are resistant to antibiotics that specifically inhibit bacterial cell wall synthesis, like the penicillins. If detected early, the diseases associated with invasive mycoplasmal infections are treatable with long cycles of high-dose antibiotics. Since the organism is a slow-growing, intracellular type with a long life cycle, several long-term courses of antibiotics may be necessary. The infection may need to be treated for several months or years. If a person is taking antibiotics, the testing will not detect the presence of Mycoplasma in the blood. And, if a person has been taking antibiotics, they must wait for 2-3 months after stopping the antibiotics for the test to be accurate.

As yet, it is not known if antibiotics are a cure for mycoplasma infections. Mycoplasma has the ability to enter any cell and alter itself, changing its cellular makeup with every cell division. This may make it impossible for readily-available antibiotics to clear the body of this organism. The recommended treatment appears to cause the organism to diminish or go dormant. Once the natural killer cells of the immune system are free of the infection, they can become healthier and keep the mycoplasma under control. This may take several months or years of antibiotic treatment to accomplish. During this time, it is important to not lower the dose or stop taking the antibiotic too early, for a relapse is certain.

Mycoplasmas: Sophisticated, Reemerging, and Burdened by Their Notoriety

No other group of procaryotes has been so embroiled in controversy and in establishing a clear pathogenic niche as the mycoplasmas. Their virulence determinants are undeniably complex, and their unique biological properties likely challenge the host differently from typical bacterial pathogens. Also, numerous *Mycoplasma* species appear to comprise the commensal microbial flora of healthy persons, and the association of these mycoplasmas with disease complicates the diagnosis and necessitates extensive and highly specific serologic, nucleic acid, and epidemiologic data. Nonetheless, mycoplasmas by themselves can cause acute and chronic diseases at multiple sites with wide-ranging complications and have been implicated as cofactors in disease. Recently, mycoplasmas have been linked as a cofactor to AIDS pathogenesis and to malignant transformation, chromosomal aberrations, the Gulf War Syndrome, and other unexplained and complex illnesses, including chronic fatigue syndrome, Crohn's disease, and various arthritides. Even with mounting evidence of their pervasive and pathogenic potential, mycoplasmas still evoke the image of a group of obscure or impotent microorganisms. Yet they are evolutionarily advanced procaryotes, and their elite status as "next generation" bacterial pathogens necessitates new paradigms in fully understanding their disease potential.

Mycoplasmas, which lack cell walls but possess distinctive sterol-containing plasma membranes, are taxonomically separated from other bacteria and belong to the class Mollicutes (*mollis*, soft; *cutis*, skin). Mollicutes, a term that includes the cell wall-less procaryotes assigned to numerous genera under the class Mollicutes and is frequently used interchangeably with mycoplasmas, are unusual for other biological reasons as well. They are evolutionary descendants of the low G+C containing gram-positive bacteria and, through chromosome reduction, represent the smallest self-replicating life forms. Their streamlined genome size, which illustrates extreme biological gene economy, imposes complex nutritional requirements, such as dependence on external supplies of biosynthetic precursors, including amino acids, nucleotides, fatty acids, and sterols. This limited coding capacity dictates for mycoplasmas a parasitic way of life that few pathogenic micro-organisms can claim. Therefore, the view that pathogenic mycoplasmas can grow "independently" requires an appreciation of their fastidious nature and their intimate dependence upon the host. Because of these properties, pathogenic mycoplasmas are among the most difficult micro-organisms to grow from clinical specimens and remain frequent contaminants of primary and continuous eucaryotic cell lines and tissue cultures. In some instances, mycoplasma contamination is obvious since infected eucaryotic cells exhibit aberrant growth, metabolism, and morphology. However, mycoplasmas often establish covert and chronic infections of target cells that lead to either invalid and misleading data or introduction of mycoplasmas or their products into reagents dedicated to therapeutic or research purposes. The recent emphasis on isolating viral agents, such as human immunodeficiency virus (HIV)-1, from human primary lymphocytic cells has also demonstrated the frequent cocultivation of mycoplasmas of human origin. Often, the unwanted sources of exogenous mycoplasmas are serum products and filter-"sterilized"(450 nm) solutions; cross-contamination by already infected cell cultures, viral stocks, or immunologic preparations; breaks in technique, including aerosols from the respiratory tract or by mouth pipetting; ignorance of the mycoplasma problem; or scientific indifference.

Detailed up-to-date reviews describing the biological and pathogenic properties of mycoplasmas have been published. Our intention here is to provide a concise historical perspective of the role of mycoplasmas in human disease; highlight the discoveries of new *Mycoplasma* species and their association with human illness and host conditions that present problems in detection and treatment; describe selected biological properties of mycoplasmas consistent with their intimate host relationship and possible mechanisms of pathogenicity; and address recent controversies associated with mycoplasmas as emerging infectious agents. Renewed attention to these issues may provide the impetus to demystify mycoplasmas and improve their standing as genuine, card-carrying pathogens.

Historical Perspectives

The earliest reports of mycoplasmas as infectious agents in humans appeared in the 1930s and 1940s. At that time, primary atypical pneumonia was associated with an infectious agent that because of its minute size and innate biological properties unknown at that time, passed through bacteria-retaining filters, resisted penicillin and sulfonamide therapies, and adapted to growth in embryonated eggs and tissue culture cells. Correlations between the etiologic agent of "walking pneumonia" with viruses, L-forms, and pleuropneumonia-like agents were frequent and often misleading. Finally, definitive studies in the early 1960s established *Mycoplasma pneumoniae* as the singular cause of cold agglutinin-associated primary atypical pneumonia. Today *M. pneumoniae* remains an important cause of pneumonia and other airway disorders, such as tracheobronchitis and pharyngitis, and is associated with extrapulmonary manifestations, such as hematopoietic, exanthematic, joint, central nervous system, liver, pancreas, and cardiovascular syndromes.

The confusion associated with *M. pneumoniae*-mediated infections has recurred many times with other mycoplasmas, whose detection in clinical specimens through culture, antibody, or DNA-based testing is frequently dismissed as "only mycoplasmas" even when they appear to be the primary pathogens. Two mycoplasmas commonly found in the urogenital tracts of healthy persons are *Mycoplasma hominis* and *Ureaplasma urealyticum*. However, over the years, the pathogenic roles of these mycoplasmas have been proven in adult urogenital tract diseases, neonatal respiratory infections, and a range of other diseases usually in immunocompromised patients.

The increasing impact of *Mycoplasma* species on emerging diseases. *Mycoplasma fermentans* strains were first isolated from the lower genital tract of both adult men and women in the early 1950s, but their role in classic lower genital tract disease has not been established. Reports in the 1970s of *M. fermentans* in the joints of rheumatoid arthritis patients and in the bone marrow of children with leukemia raised expectations for its pathogenic potential; these findings have not been adequately confirmed. Sufficient evidence, however, has accumulated recently to establish an important and emerging role for *M. fermentans* in human respiratory and joint diseases. *M. fermentans* has been detected by specific gene amplification techniques such as polymerase chain reaction (PCR) in the synovial fluid of patients with inflammatory arthritis, but not in the joints of patients with juvenile or reactive arthritis. In two other studies using PCR, *M. fermentans* was identified in the upper respiratory tract of 20 per cent to 44 per cent of both healthy and HIV-infected patients and was associated with acute respiratory distress syndrome in nonimmunocompromised persons.

Mycoplasma genitalium was detected in the urogenital tract of two patients with nongonococcal urethritis in 1981, but for more than a decade, very little was known about its host distribution and pathogenicity. Early experimental studies established that the organism caused lower genital tract infections in both male and female chimpanzees, with extensive urethral colonization in males and apparent tissue invasion, eventually leading to overt bacteremia. However, the fastidious growth requirements of *M. genitalium* from human hosts severely limited further study until the advent of molecular detection techniques. Specific sequences in the 140 kDa adhesin protein gene of *M. genitalium* were selected as targets in a PCR-based detection assay. Subsequent application of these techniques in cases of acute nongonococcal urethritis, not including those of patients colonized or infected with *Chlamydia trachomatis,* has provided mounting evidence for the involvement of *M. genitalium* as an etiologic agent of this disease. Also, *M. genitalium* has been suspected in chronic nongonococcal urethritis and pelvic inflammatory disease.

The discovery in 1988 of *M. genitalium* strains in human nasopharyngeal throat specimens, where they were frequently mixed with strains of *M. pneumoniae*, not only changed dramatically the concept of host distribution of *M. genitalium* but also prompted critical questions about the role of this mycoplasma in human respiratory disease. However, the immunologic cross-reactivity between *M. genitalium* and *M. pneumoniae* and the inability of most conventional diagnostic serologic tests to conclusively identify *M. genitalium* have complicated its delineation in acute human respiratory disease. PCR assays specific for the organism have detected *M. genitalium* in throat specimens of patients infected with HIV-1. However, these probes have not been applied to control groups and patients in outbreaks of acute respiratory disease and/or pneumonia to determine whether *M. genitalium* alone is an etiologic agent in respiratory infections.

M. genitalium has been implicated as an etiologic agent in certain human joint diseases. This clinical correlation began with the observation of a mixed infection of *M. pneumoniae* and *M. genitalium* in synovial fluid specimens of a nonimmunocompromised patient after an acute respiratory infection. A predominant role was not established for either

Mycoplasma species in the initial respiratory disease or in the joint manifestations, although evidence to implicate postinfectious autoimmunity to both organisms was described. These findings prompted a PCR assay on synovial fluids from patients with various arthritic syndromes, which presented case reports on two of 13 patients with *M. genitalium* detected in joint fluids.

Another area of emerging mycoplasmal infections concerns immunodeficiency. Although patients with congenital or acquired disorders of antibody production are susceptible to a wide variety of microbial infections, the unique susceptibility of such patients to mycoplasmal infections is a growing concern, especially considering the number of occurrences, the types of mycoplasmas involved, and the difficulties posed in the therapeutic management of such infections. In addition, the increased use of prolonged or permanent immunosuppressive chemotherapy required for patients undergoing tissue or organ transplantation or treatment of various malignant diseases has also increased the risk for mycoplasmal infections from mycoplasmas that are part of the normal human mollicute flora to those acquired through animal contact.

The association between immunodeficiency and mycoplasmal infections was first reported in the mid 1970s in patients with primary hypogammaglobulinemia and infection with *U. urealyticum, M. pneumoniae, Mycoplasma salivarium,* and *M. hominis* that localized in joint tissue, frequently with destructive arthritis. Similar joint infections in hypogammaglobulinemic patients with these mycoplasmal species continue to be reported. Since most of these mollicutes, with the possible exception of *M. pneumoniae*, occur as part of the normal human flora, the origin of such joint infections is considered endogenous. Patients with hypogammaglobulinemia and other antibody deficiencies are also especially susceptible to mycoplasmal infections of the upper respiratory and urinary tracts caused most frequently by *M. pneumoniae* or *U. urealyticum*, respectively.

Mycoplasmal infections following organ transplantation and immunosuppressive chemotherapy were observed in the early 1980s, with both *M. hominis* and *U. urealyticum* reported most often. Although these infections most likely originated from the patient's normal microbial flora, a recent report of donor transmission of *M. hominis* to two lung allograft recipients suggests that donor tissue may be a more important factor in transplant infections than currently recognized.

While patients with antibody defects or those receiving immunosuppressive drugs appear to be the most susceptible to infections with mycoplasmas present in healthy tissues, emerging evidence indicates that contact with other mycoplasmas in the environment is an important hazard. The direct isolation of a feline mycoplasma (*M. felis*) from the joint of a hypogammaglobulinemic patient with septic arthritis was recently reported, with suspected transmission occurring through a cat bite 6 months before the onset of arthritis. Fatal septicemia caused by *M. arginini*, a common animal mycoplasma, from blood and multiple tissue sites in a slaughter house employee who had advanced non-Hodgkin's lymphoma and hypogammaglobulinemia, and a septicemic infection with a canine mycoplasma (*M. edwardii*) in a patient with advanced AIDS.

One of the most critical aspects of mycoplasmal infections in immunodeficient patients is the frequent inability to control such infections with appropriate broad spectrum antibiotics. Although the tetracyclines and erythromycins are effective chemotherapeutic agents for many mycoplasmal infections, *M. fermentans* and *M. hominis* strains are usually resistant to erythromycin, and tetracycline-resistant strains of *M. hominis* and*U. urealyticum* have been reported from the lower urogenital tract of patients. However, these antibiotics and most other broad spectrum agents have limited mycoplasmacidal activity in vivo, and their efficacy eventually depends on an intact host immune system to eliminate the mycoplasmas. Most hypogammaglobulinemic patients lack the ability to mount a strong antibody response. Guidelines for managing such mycoplasmal infections in patients with immune defects should include immediate in vitro testing of the isolated mollicute against a wide range of antibiotics; expeditious administration of the antibiotic by the most appropriate route (intravenously, if warranted); prolonged therapy terminated only if there is no rapid clinical or microbiological response; and possibly administration of intravenous immunoglobulin. Clinical management of mycoplasmal infections in transplant patients is more difficult since immunoglobulins may enhance graft or organ rejection. In the absence of suitable mycoplasmacidal chemotherapeutic agents, vigourous and sustained chemotherapy with the most active antibiotic is the current method of choice.

Mechanisms of Pathogenicity

Many mycoplasmal pathogens exhibit filamentous or flask-shaped appearances and display prominent and specialized polar tip organelles that mediate attachment to host target cells. These tip structures are complex, composed of a network of interactive proteins, designated adhesins, and adherence-accessory proteins. These proteins cooperate structurally and functionally to mobilize and concentrate adhesins at the tip and permit mycoplasmal colonization of mucous membranes and eucaryotic cell surfaces, probably through host sialoglycoconjugates and sulfated glycolipids. It appears that mycoplasmal cytadherence-related proteins represent a superfamily of genes and proteins that have been conserved through horizontal gene transfer from an ancestral gene family. This protein network resembles a specialized cytoskeleton-like apparatus, which may represent the precursor to mammalian cytoskeletal and extracellular matrix-like complexes. Other *Mycoplasma* species lack distinct tip structures yet are capable of cytadherence, and they may use related genes or proteins or alternative mechanisms of surface parasitism.

The family of mycoplasmal genes and proteins involved in cytadherence has been studied most extensively in *M. pneumoniae*. Noncytadhering phenotypes that arise through spontaneous mutation at high frequency have been categorized into mutant classes on the basis of distinct protein profiles. These noncytadhering mycoplasmas cannot synthesize specific cytadherence-related proteins or are unable to stabilize them at the tip organelle, which leads to abnormal anatomical tip structures and avirulence. Spontaneous reversion to the cytadhering phenotype is accompanied by the reappearance of the implicated proteins, restoration of structurally and functionally intact tips, and return of full infectivity. Similar cytadherence-related genes and proteins have been reported for *M. genitalium* on the basis of biochemical, immunologic, and genetic analyses. Furthermore, striking similarities exist in the order of operons that comprise the cytadherence-related genes and the organization of these genes within each operon of *M. pneumoniae* and *M. genitalium*. These similarities reinforce the unexpected coisolations of *M. genitalium*, along with *M. pneumoniae*, from the nasopharyngeal throat swabs of patients with acute respiratory diseases and from synovial fluids of patients with arthritis as described in the previous section. The isolation of *M. pneumoniae* from the human urogenital tract further suggests that these mycoplasmas have evolved parasitic strategies that include overlapping tissue tropisms as determined by the genetic and chemical relatedness of their cytadherence genes and proteins. The recent use of transposon mutagenesis to generate *M. pneumoniae* and *M. genitalium* transformants displaying cytadherence-deficient phenotypes should further clarify the relationships between the cytadherence-related genes and proteins and identify additional sites previously unlinked to cytadherence.

An interesting feature of specific *M. pneumoniae* and *M. genitalium* adhesins is their multiple gene copy nature. Although only one full-length copy of the adhesin structural genes exists in adhesin-related operons, precise regions of these adhesin genes are detected as single genomic copies, while other regions occur as closely homologous, but not identical, multiple copies. In other words, multiple truncated and sequence-related copies of the adhesin genes are dispersed throughout the genome, which could generate adhesin variation through homologous recombination. Consistent with this possibility is the existence of restriction fragment length polymorphisms in the adhesin genes of human clinical isolates of *M. pneumoniae* and *M. genitalium*, reflected by sequence divergence in the multiple-copy regions of the adhesin genes. It appears that a repertoire of partial adhesin-related gene regions serves as a reservoir to regulate the structural and functional properties of mycoplasmal adhesins through recombination events, which may lead to circumvention of the host immune response. Mechanisms of phase and antigenic variation are likely to occur in which mycoplasmal adhesins exhibit altered specificities and affinities, as determined by the organization of constant and variable adhesin gene sequences. Therefore, despite their small genomes, pathogenic mycoplasmas facilitate DNA rearrangements through repetitive gene sequences, thus promoting genetic diversity and maximizing the coding potential of their limited genomes. The immunodominant epitopes of the mycoplasmal adhesins appear not to be identical to the adherence-mediating domains. The latter are in part encoded by single copy regions of the adhesin genes and are highly conserved, which reinforces their essential role in mycoplasmal recognition of host cell receptors and colonization. Host immunoresponsiveness directed at the noncytadherence-mediating variable regions is unlikely to generate effective cytadherence-blocking antibodies, which may in part clarify the observed high reinfection rates of patients. Thus, the grouping of clinical isolates of *M. pneumoniae* into two categories, on the basis of sequence divergence in the multiple-copy regions of the adhesin gene, along with the immune status of the population, may explain the epidemiologic patterns of *M. pneumoniae* reported over the years.

Another characteristic of the cytadherence-related proteins is their proline-rich composition, which markedly influences protein folding and binding. Several reports have established the importance of these proline-rich domains in mycoplasmal cytadherence and virulence, and recent evidence further suggests that mycoplasmal peptidylprolyl isomerases, i.e., cyclophilins, are critical in regulating the conformation and function of the mycoplasmal cytadherence-related tip organelle, colony morphology, and growth. In addition to this proline-rich property, one of the most unusual features of the adhesins is their extensive sequence homology to mammalian structural proteins.

This molecular mimicry is especially interesting since it has been suggested for decades that mycoplasmas provoke an antiself response that triggers immune disorders, although the basis for the induction has been elusive. Patients with documented *M. pneumoniae* respiratory infections demonstrate seroconversion to myosin, keratin, and fibrinogen and exhibit extrapulmonary manifestations, such as exanthems and cardiac abnormalities.

Bacteria-mediated autoimmune disorders is the development of acute rheumatic fever following streptococcal infection. Antistreptococcal antibodies reactive against ahelical coiled-coil regions of the M protein crossreact with heart myosin, tropomyosin, and mycoplasmal adhesins. In the latter case, these mycoplasmal adhesins exhibit amino acid sequence homologies with human CD4 and class II major histocompatibility complex lymphocyte proteins, which could generate autoreactive antibodies and trigger cell killing and immunosuppression. Also, mycoplasmas may serve as B-cell and T-cell mitogens and induce autoimmune disease through the activation of antiself T cells or polyclonal B cells.

The multiorgan protean manifestations of mycoplasmal infections in humans are consistent with the pathogenesis of autoimmunity. Furthermore, the ability of mycoplasmas to induce a broad range of immunoregulatory events, mediated by cytokine production and direct effects on macrophages, B and T cells, and glial cells, is evidence that mycoplasmas possess the attributes of primary mediators of pathogenesis. Cytokine production and lymphocyte activation may either minimize disease through the activation of host defence mechanisms or exacerbate disease through lesion development. Also, a superantigen derived from *Mycoplasma arthritidis*, a mycoplasma pathogenic for rodents, induces arthritis and chronic disease manifestations. It has been suggested that related superantigen-like molecules may exist in mycoplasmas of human origin triggering autoimmune and other inflammatory pathologies.

It appears that cytadherence is the initial step in the virulence process of pathogenic mycoplasmas and precedes a spectrum of subtle or overt host cell responses. In specific instances, distinct cytopathology correlates with the infecting *Mycoplasma* species, the number of adherent mycoplasmas, the length of coincubation, the induction of proinflammatory cytokines, and the age and immune status of the patient. The exacerbation of clinical syndromes may correlate with a history of mycoplasmal infection as observed in patients with recurrent *M. pneumoniae* exposures. Also, the elevated expression of proinflammatory cytokines associated with mycoplasmal disease pathogenesis may coincide with the intensity of the symptoms. In other cases, chronic disease or no obvious signs or symptoms of disease accompany mycoplasmal infection.

Other biological properties of mycoplasmas have been implicated as virulence determinants and include

1. Generation of hydrogen peroxide and superoxide radicals by adhering mycoplasmas, which induces oxidative stress, including host cell membrane damage;
2. Competition for and depletion of nutrients or biosynthetic precursors by mycoplasmas, which disrupts host cell maintenance and function;
3. Existence of capsule-like material and electron-dense surface layers or structures, which provides increased integrity to the mycoplasma surface and confers immunoregulatory activities;
4. High-frequency phase and antigenic variation, which results in surface diversity and possible avoidance of protective host immune defences;
5. Secretion or introduction of mycoplasmal enzymes, such as phospholipases, ATPases, hemolysins,

proteases, and nucleases into the host cell milieu, which leads to localized tissue disruption and disorganization and chromosomal aberrations; and

6. Intracellular residence, which sequesters mycoplasmas, establishes latent or chronic states, and circumvents mycoplasmicidal immune mechanisms and selective drug therapies. Whether pathogenic mycoplasmas enter and survive within mammalian cells has been debated for many years.

Consistent with this possibility, mycoplasmas exhibit limited biosynthetic capabilities; are highly fastidious and dependent upon the host microenvironment and complex culture medium for growth; have been observed in intimate contact with mammalian cell surfaces and within target cells; may be capable of initiating fusion with host cells through their cholesterol-containing unit membranes; and survive long-term recommended antimicrobial treatment in humans and tissue cultures. Recent sightings of intact mycoplasmas throughout the cytoplasm and the perinuclear regions of tissue cells from infected patients and in cell cultures, along with evidence that mycoplasmas are capable of long-term intracellular survival and replication in vitro, offer an additional dimension to the pathogenic potential of mycoplasmas.

The Latest Controversies: Food for thought or the Twilight Zone

On the basis of the above information, the virulence strategies displayed by mycoplasmas are likely the summation of a multitude of biological activities. Since no obvious single or group of mycoplasmal properties inextricably correlates with disease manifestations, the proof that mycoplasmas are card-carrying pathogens necessitates thorough and highly specific microbiological, epidemiologic, and diagnostic criteria; detailed descriptions of biochemical, genetic, and immunologic characteristics that distinguish virulent and avirulent mycoplasmas; and reproducibility of the symptoms of disease in experimental animal models or in the natural spread of

infection among susceptible populations. The portfolio of available evidence concerning mycoplasma-mediated disease pathogenesis is limited. These scientific shortcomings precipitate misconceptions concerning mycoplasmas as singular agents of infectious diseases, as putative cofactors in the progression of other diseases, and as universal contaminants of cell cultures. Clearly, multiple pathways of interaction with target cells appears to be the *modus operandi* of the *Mycoplasma* species. With this conceptual scientific framework as a background, five recently proposed and controversial associations of mycoplasmas to human diseases are worth noting.

AIDS

The role of mycoplasmas in accelerating the progression of AIDS could not have begun under more baffling and circuitous conditions. A virus-like agent that arose through transfection of NIH 3T3 cells with DNA from Kaposi sarcoma tissues of AIDS patients was later shown to be *M. fermentans*. The spotted history of *M. fermentans* in rheumatoid arthritis and leukemia and its frequent contamination of cell cultures, along with its contemporary link to AIDS, have been considerable impediments to overcome in its elevation to pathogenic status. However, careful and convincing independent studies by several laboratories have implicated *M. fermentans* as a cause of systemic infections and organ failure in AIDS patients.

The isolation of *M. fermentans* from blood and urine samples of HIV-infected persons, its detection by PCR and immunohistochemistry in multiple tissue sites at various stages of AIDS, and its ability to stimulate CD4+ lymphocytes and other immunomodulatory activities implicate this *Mycoplasma* species as a cofactor in AIDS. Consistent with this possibility, *M. fermentans* has been shown to act synergistically with HIV to enhance cytopathic effects on human CD4+ lymphocytes. Coincident with these studies, a new *Mycoplasma* species, *Mycoplasma penetrans*, also has emerged as a potential cofactor in AIDS progression. Its isolation almost exclusively from the urine of HIV-infected patients, the extraordinarily high prevalence of antibodies against this mycoplasma in HIV-infected patients and not in HIV-seronegative persons, and its capacity to invade target cells and activate the immune system of HIV-infected patients at various stages of disease correlate with a synergistic role with HIV. Other mycoplasmas, including *M. genitalium* and *Mycoplasma pirum*, have also been isolated from AIDS patients and implicated as potential cofactors.

However, the proposed role of mycoplasmas as infectious agents and cofactors in AIDS-related disorders still remains a hypothesis without definitive proof. If cofactors of HIV are essential to the development of late stages of HIV-mediated disease, mycoplasmas possess all the prerequisite properties of the consummate helper. Their ability to establish covert or overt chronic and persistent infections with concomitant activation of the immune system, stimulation of cytokine production, and induction of oxidative stress correlate with increased HIV replication and disease progression. Are mycoplasmas irrelevant to AIDS, or are the clinical and microbiological correlations sufficient to imply intimate relationships between HIV and mycoplasmas, especially as the infected host undergoes immunologic distress?

Malignant Transformation

As early as the mid-1960s, mycoplasma-infected cell lines were associated with chromosomal aberrations, altered morphologies, and cell transformation. These abnormal oncogenic cell traits continued even after the apparent elimination of mycoplasmas, and evidence implied increased tumorigenicity of these transformed cells in animals. This issue has been revisited in studies demonstrating that longterm, persistent mycoplasmal infection of mouse embryo cells initiated a multistage cellular process that resulted in irreversible cell transformation, karyotypic alterations, and tumorigenicity in nude mice.

Gulf War Syndrome

One of the most controversial current medical issues is whether the multiple acute and chronic symptoms found in veterans of the Persian Gulf War were caused by chemical exposure, infectious agents, or psychological problems, or whether a Gulf War Syndrome exists at all. The clinical illness comprises a collection of symptoms, including chronic fatigue, joint pain, headaches, and skin rashes. One study suggests that pathogenic mycoplasmas are responsible for a large number of cases among veterans, on the basis of DNA hybridization and the responsiveness of veterans to prolonged antibiotic treatment. Even though the experimental evidence is sparse and incomplete and well-controlled and detailed studies by independent laboratories are needed, if the Gulf War Syndrome has infectious causes, mycoplasmas with their requisite biological credentials are potential candidates.

Crohn's Disease

Several epidemiologic studies correlate respiratory infections with exacerbation of Crohn's disease and other chronic inflammatory bowel diseases. Acute onset gastrointestinal symptoms in patients with these diseases are accompanied by seroconversion to specific viral or *M. pneumoniae* antigens. As indicated earlier, mycoplasmas can elicit pleiotropic immune responses and are difficult to eliminate in patients despite appropriate antibiotic treatment. Steroid therapy to control gastrointestinal symptoms in these patients, along with the multifaceted biological properties associated with pathogenic mycoplasmas, may precipitate the onset of acute exacerbations of chronic inflammatory bowel disease.

Rheumatoid Arthritis and Other Human Arthritides

The occurrence of various *Mycoplasma* and *Ureaplasma* species in joint tissues of patients with rheumatoid arthritis, sexually transmitted reactive arthritis, and other human arthritides can no longer be ignored. A clinical trial of longterm

(6 to 12 months) antibiotic (doxycycline) therapy before cartilage destruction might prove beneficial in managing such frequent and often debilitating infections.

Extensive clinical and microbiological evidence indicates that mycoplasmas alone can elicit a spectrum of illness for which no other agents are incriminated. The eradication of these pathogenic mycoplasmas from various tissue sites requires an intact and functional immune system, although persons with fully competent immune systems may have difficulty eliminating mycoplasmas, even with recommended prolonged drug therapy. Nonetheless, mycoplasmas are still viewed as subordinates to other infectious agents and are relegated to a category of commensals that unwittingly cause disease in patients whose immune systems offer little resistance to microbial stress and overload.

The fundamental importance of mycoplasmas in specific diseases of humans, animals, insects, and plants is irrefutable, and their unique biological properties are consistent with their intimate association with host target cells. These remarkable bacteria must continue to receive the scientific attention of mycoplasmologists, cell culturists, clinicians, immunologists, and DNA sequencers who most recently are compiling extensive databases that may eventually dissect every approachable mycoplasmal element that defines their biological and genetic being. Nonetheless, mycoplasmas remain mysterious and enigmatic, and the available data and proposed hypotheses that correlate mycoplasmas with disease pathogenesis range from definitive, provocative, and titillating to inconclusive, confusing, and heretical. Controversy seems to be a recurrent companion of mycoplasmas, yet good science and openmindedness should overcome the legacy that has burdened them for decades.

5

Rickettsiae

Rickettsiae are small, Gram-negative bacilli that have evolved in such close association with arthropod hosts that they are adapted to survive within the host cells. They represent a rather diverse collection of bacteria, and therefore listing characteristics that apply to the entire group is difficult. The common threads that hold the rickettsiae into a group are their epidemiology, their obligate intracellular lifestyle, and the laboratory technology required to work with them. In the laboratory, rickettsiae cannot be cultivated on agar plates or in broth, but only in viable eukaryotic host cells (e.g., in cell culture, embryonated eggs, or susceptible animals). The exception, which shows the artificial nature of using obligate intracellular parasitism as a defining phenotypic characteristic, is *Bartonella (Rochalimaea) quintana*, which is cultivable axenically, but was traditionally considered as a rickettsia. The diversity of rickettsiae is demonstrated in the variety of specific intracellular locations where they live and the remarkable differences in their major outer membrane proteins and genetic relatedness. An example of extreme adaptation is that the metabolic activity of *Coxiella burnetii* is greatly increased in the acidic environment of the phagolysosome, which is a harsh location for survival for most other organisms. Obligate intracellular parasitism among bacteria is not unique to rickettsiae. Chlamydiae also have evolved to occupy an intracellular niche, and numerous bacteria (e.g., *Mycobacteria, Legionella, Salmonella, Shigella, Francisella,* and *Brucella*) are facultative intracellular parasites. In contrast with chlamydiae, all rickettsiae can synthesize ATP. *Coxiella burnetii* is the only rickettsia that appears to have a developmental cycle.

DESCRIPTION AND SIGNIFICANCE

Rickettsia bacteria are well known pathogens. *Rickettsia conorii* causes Mediterranean spotted fever in humans and is contracted by contact with infected brown dog ticks. Other Rickettsia include *Rickettsia prowazekii,* which causes typhus, *R. rickettsii,* which causes Rocky Mountain spotted fever, and *Rickettsia akari,* which causes rickettsialpox. In addition to this, the genome of *Rickettsia prowazekii* is similar to mitochondrial genomes; phylogenetically, *R. prowazekii* is more closely related to the mitochondria than any other microbe known thus far.

Genome Structure

The genome of *Rickettsia prowazekii* is 1,111,523 base pairs in length and contains 834 protein-coding genes. It contains no genes for anaerobic glycolysis as well as genes involved in the biosynthesis and regulation of biosynthesis of amino acids and nucleosides in free-living bacteria similar to mitochondrial genomes. Unlike the mitochondrial genome, however, the genome of *R. prowazekii* contains a complete set of genes encoding for the tricarboxylic acid cycle and the respiratory-chain complex. Still, the genomes of *Rickettsia* as well as the mitochondria are small, highly derived, "products of several types of reductive evolution".

CELL STRUCTURE AND METABOLISM

Rickettsia bacteria are obligate intracellular pathogens that are dependent on entry, growth, and replication within the cytoplasm of a eukaryotic host cell. The host cell then lysis and releases the rickettsial progeny to initiate a new infection cycle. The infection generally doesn't result in complete shutdown of the host machinery. Apparently, "vigourous host responses" generally clear the rickettsial pathogens. Conversely, the host's immune responses can also lead to the persistence of a subclinical infection even years past primary infection and/or antibiotic treatment. One theory on how rickettsiae survives in host cells has to do with the "suppression of the antimicrobial activities of the eukaryotic target cells, specifically monocytes/macrophages".

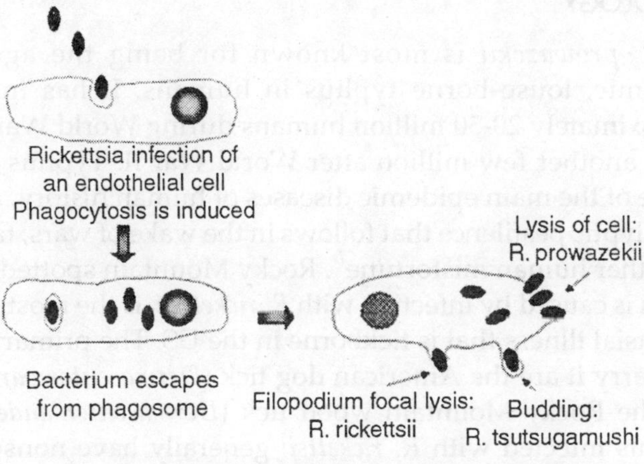

Fig. Rickettsial infection of endothelial cells.

ECOLOGY

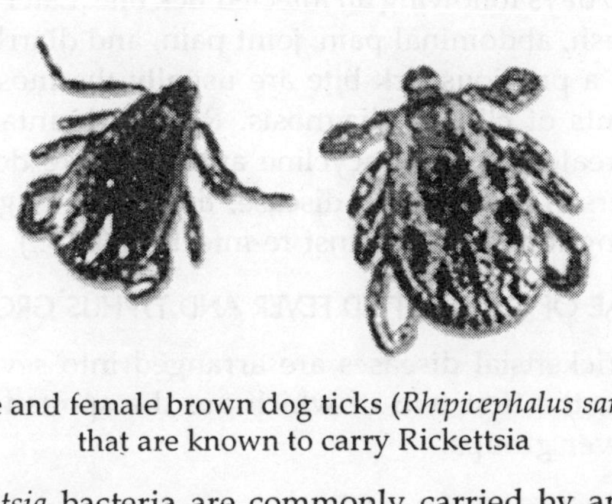

Fig. Male and female brown dog ticks (*Rhipicephalus sanquineus*) that are known to carry Rickettsia

Rickettsia bacteria are commonly carried by anthropods like ticks, mites, lice, or fleas. Another way for animals and humans to contract the bacteria is through wild rodents that have been infected with a *Rickettsia* bacteria by these louse. Different forms of *Rickettsia* and the diseases that they cause can be found all over the world. While some diseases themselves are worldwide, others, such as Oriental spotted fever caused by *R. japonica* in Japan, are localized to one general place or area.

PATHOLOGY

R. prowazekii is most known for being the agent of epidemic, louse-borne typhus in humans. It has infected approximately 20-30 million humans during World War I and killed another few million after World War II. Typhus 'ranks as one of the main epidemic diseases of human history, a truly apocalyptic pestilence that follows in the wake of wars, famine, and other human misfortune". Rocky Mountain spotted fever, which is caused by infection with *R. rickettsii*, is the most severe rickettsial illness that is tickborne in the US. The primary ticks that carry it are the American dog tick (*Dermacentor variabilis*) and the Rocky Mountain wood tick (*Dermacentor andersoni*). Patients infected with *R. rickettsii* generally have nonspecific symptoms including fever, nausea, vomiting, muscle pain, lack of appetite, and severe headache after an incubation period about 5-10 days following an infected tick bite. Later symptoms include rash, abdominal pain, joint pain, and diarrhea. Fever, rash, and a previous tick bite are usually the most common components of clinical diagnosis. Rocky Mountain spotted fever is treated by a tetracycline antibiotic like doxycycline; once a person has had the disease, they are thought to have long lasting immunity against re-infection (CDC).

RICKETTSIAE OF THE SPOTTED FEVER AND TYPHUS GROUPS

The rickettsial diseases are arranged into several major categories, the first two of which are the spotted fever and typhus fever groups.

Clinical Manifestations

Rocky Mountain spotted fever

Rocky Mountain spotted fever is among the most severe of human infectious diseases, with a mortality of 20 to 25 per cent unless treated with an appropriate antibiotic. The severity and mortality are greater for men, elderly persons, and black men with glucose-6-phosphate dehydrogenase deficiency. Although, in theory, the disease is always curable by early, appropriate treatment, the case fatality rate is still 4 per cent. The incidence of disease parallels the geographic distribution of infected *Dermacentor variabilis* ticks in the eastern United States and *D andersoni* in the Rocky Mountain states, where the infection was first recognized.

Rocky Mountain spotted fever was subsequently recognized in the eastern United States. The incidence has declined in the Rocky Mountain states and increased dramatically in the southeastern United States and Oklahoma. Currently most cases actually occur in the Atlantic states from Maryland to Georgia, as well as in Oklahoma, Missouri, Kansas, Ohio, Tennessee, Arkansas, and Texas, although cases are reported in nearly every state. In the southeastern states, the disease occurs during the seasonal activity of *D variabilis* ticks and affects children more frequently than adults. Significant changes in incidence do occur. From a low of 199 cases reported in 1959, the annual number of cases rose steadily to a peak of 1,192 cases in 1981, with a subsequent decline and plateau of approximately 700 cases since 1985. The reasons for these fluctuations are unclear.

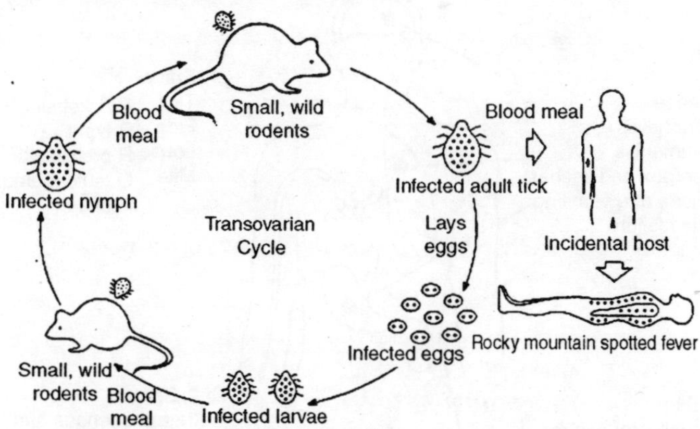

Fig. Transovarian passage of *R rickettsii* in the tick vector is an important cycle in maintaining the infection in nature from one generation of tick to another.

The rickettsiae are maintained in nature principally by transovarial transmission from infected female ticks to infected ova that hatch into infected larval offspring. A low rate of acquisition of rickettsiae by uninfected ticks occurs when the ticks feed upon small mammals with enough rickettsiae in their blood to establish tick infection. This effect replenishes lines of infected ticks that are occasionally killed by massive rickettsial overgrowth. A recently observed factor of potential importance in this balance of nature is the interference phenomenon, by which infection of ticks with nonpathogenic spotted fever group rickettsiae prevents the establishment of infection by *R rickettsii*.

The clinical gravity of Rocky Mountain spotted fever is due to severe damage to blood vessels by *R rickettsii*. This organism is unusual among rickettsiae in its ability to spread and invade vascular smooth muscle cells as well as endothelium. Damage to the blood vessels in the skin in locations of the rash leads to visible hemorrhages in one-half of all infected persons. Attempted plugging of vascular wall destruction consumes platelets, with consequent thrombocytopenia also affecting approximately one-half of the patients.

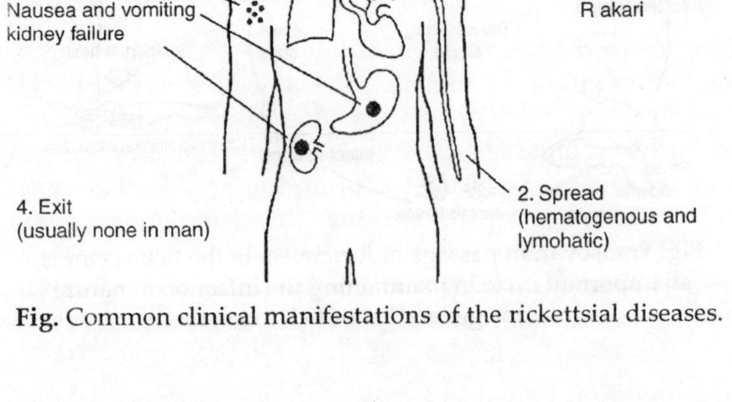

Fig. Common clinical manifestations of the rickettsial diseases.

Rickettsialpox and Other Spotted Fevers

In the 1940s an epidemic of disease characterized by fever, rash, and cutaneous necrosis appeared in one area of New York City. The etiology was traced to *R akari* transmitted by the bite of mites (*Liponyssoides sanguineus*) that infested the numerous mice in an apartment house in this area. The disease was named rickettsialpox because many patients had blister-like rashes resembling those of chickenpox. Epidemics were diagnosed in other cities, and *R akari* has been isolated in other countries. Perhaps because this nonfatal disease is seldom considered by physicians, or its incidence is truly low, the diagnosis is rarely made. Transovarial transmission in the mite and periodic documentation of cases assure us that the etiologic agent is still with us.

Boutonneuse fever, so called because of the papular rash in some cases, has many synonyms, reflecting different geographic regions of occurrence (e.g., Mediterranean spotted fever, Kenya tick typhus, and South African tick bite fever). Cases are observed in the United States in travellers returning from endemic areas. The agent, *R conorii,* is closely related to *R. rickettsii*. Severe disease resembling Rocky Mountain spotted fever can cause death in high-risk groups (e.g., elderly, alcoholic, and glucose-6-phosphate dehydrogenase-deficient patients). Cutaneous necrosis caused by rickettsial vascular infection at the tick bite site of inoculation, known as an eschar or tache noire, is observed in only half the patients with boutonneuse fever. The curiously high prevalence of antibodies reactive with *R conorii* in healthy populations in endemic regions might be explained by missed diagnosis of prior illness, subclinical infection, infection with an antigenically related but less pathogenic rickettsia, or nonspecificity of the laboratory test.

Other spotted fevers occur in geographic distributions of little concern to many physicians in the United States. North Asian tick typhus caused by *R sibirica,* Queensland tick typhus caused by *R australis,* and the recently discovered oriental spotted fever caused by *R japonica* demonstrate that spotted fever group rickettsiae occur worldwide.

Epidemic Typhus and Brill-Zinsser Disease

Epidemics of louse-borne typhus fever have had important effects on the course of history; typhus in one army but not in the opposing force has determined the outcome of wars. Populations have been decimated by epidemic typhus. During and immediately after World War I, 30 million cases occurred, with 3 million deaths. Unsanitary, crowded conditions in the wake of war, famine, flood, and other disasters and in poor countries today encourage human louse infestation and transmission of *R prowazekii*. Epidemics usually occur in cold months in poor highland areas, such as the Andes, Himalayas, Mexico, Central America, and Africa. Lice live in clothing, attach to the human host several times daily to take a blood meal, and become infected with *R prowazekii* if the host has rickettsiae circulating in the blood. If the infected louse infests another person, rickettsiae are deposited on the skin via the louse feces or in the crushed body of a louse. Scratching inoculates rickettsiae into the skin.

Between epidemics *R prowazekii* persists as a latent human infection. Years later, when immunity is diminished, some persons suffer recrudescent typhus fever (Brill-Zinsser disease). These milder sporadic cases can ignite further epidemics in a susceptible louse-infested population. In the United States Brill-Zinsser disease is seen in immigrants who suffered typhus fever before entering the country. In the eastern United States, sporadic human cases of *R prowazekii* infection have been traced to a zoonotic cycle involving flying squirrels and their own species of lice and fleas.

Murine Typhus

Murine typhus is prevalent throughout the world, particularly in ports, countries with warm climates, and other locations where rat populations are high. *Rickettsia typhi* is associated with rats and fleas, particularly the oriental rat flea, although other ecologic cycles (e.g., opossums and cat fleas) have been implicated. Fleas are infected by transovarian transmission or by feeding on an animal with rickettsiae circulating in the blood. Rickettsiae are shed from fleas in the feces, from which humans acquire the infection through the skin, respiratory tract, or conjunctiva. During the 1940s more than 4,000 cases of murine typhus occurred annually in the United States. The incidence declined coincident with increased utilization of the insecticide DDT. Although the infection and clinical involvement affects the brain, lungs, and other visceral organs in addition to the skin, mortality in humans is less than 1 per cent.

Structure, Classification, and Antigenic Types

Rickettsia species include two antigenically defined groups that are closely related genetically but differ in their surface-exposed protein and lipopolysaccharide antigens. These are the spotted fever and typhus groups. The organisms in these groups are smaller (0.3 μm by 1.0 μm) than most Gram-negative bacilli that live in the extracellular environment. They are surrounded by a poorly characterized structure that is observed as an electron-lucent zone by transmission electron microscopy and is considered to represent a polysaccharide-rich slime layer or capsule.

The cell wall contains lipopolysaccharides, a major component that differs antigenically between the typhus group and the spotted fever group. These rickettsiae also contain major outer membrane proteins with both cross-reactive antigens and surface-exposed epitopes that are species specific and easily denatured by temperatures above 54°C. The major outer membrane protein of typhus group rickettsiae has an apparent molecular mass of 120,000 Da. Spotted fever group rickettsiae generally have a pair of analogous proteins with some diversity of their molecular masses.

Rickettsia prowazekii has a transport mechanism that exchanges ATP for ADP in its intracellular environment, thus providing a means to usurp host cell energy sources under favourable circumstances. Rickettsiae also are able to synthesize ATP via metabolism of glutamate. Adaptation to the intracellular environment is further evidenced in a variety of transport mechanisms to obtain crucial substances such as particular amino acids from cytoplasmic pools in the host cell. These adaptations and the presence of numerous independent metabolic activities demonstrate that rickettsiae are not degenerate forms of bacteria, but rather have evolved successfully for survival with an intracellular life-style.

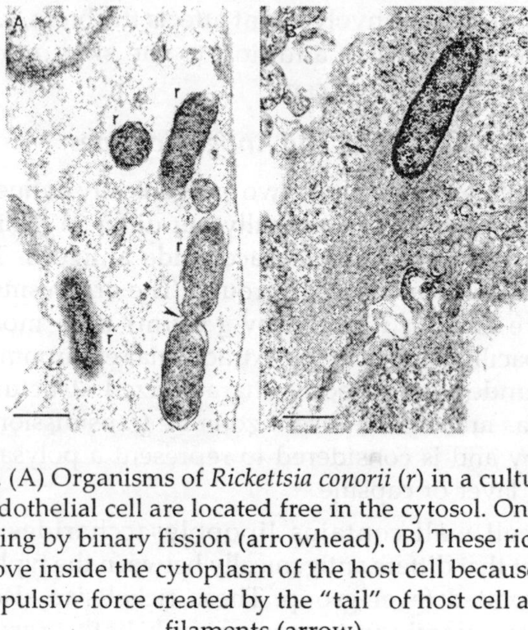

Fig. (A) Organisms of *Rickettsia conorii* (*r*) in a cultured human endothelial cell are located free in the cytosol. One rickettsia is dividing by binary fission (arrowhead). (B) These rickettsiae can move inside the cytoplasm of the host cell because of the propulsive force created by the "tail" of host cell actin filaments (arrow).

PATHOGENESIS

Rickettsiae are transmitted to humans by the bite of infected ticks and mites and by the feces of infected lice and fleas. They enter via the skin and spread through the bloodstream to infect vascular endothelium in the skin, brain, lungs, heart, kidneys, liver, gastrointestinal tract, and other organs. Rickettsial attachment to the endothelial cell membrane induces phagocytosis, soon followed by escape from the phagosome into the cytosol. Rickettsiae divide inside the cell. *Rickettsia prowazekii* remains inside the apparently healthy host cell until massive quantities of intracellular rickettsiae accumulate and the host cell bursts, releasing the organisms. In contrast, *R rickettsii* leaves the host cell via long, thin cell projections (filopodia) after a few cycles of binary fission. Hence, relatively few *R rickettsii* organisms accumulate inside any particular cell, and rickettsial infection spreads rapidly to involve many other cells. Perhaps because of the numerous times the host cell membrane is traversed, there is an influx of water that is initially sequestered in cisternae of cytopathically dilated rough endoplasmic reticulum in the cells more heavily infected with *R rickettsii*.

The bursting of endothelial cells infected with *R prowazekii* is a dramatic pathologic event. The mechanism is unclear, although phospholipase activity, possibly of rickettsial origin, has been suggested. Injury to endothelium and vascular smooth muscle cells infected by *R rickettsii* seems to be caused directly by the rickettsiae, possibly through the activity of a rickettsial phospholipase or rickettsial protease or through free-radical peroxidation of host cell membranes. Host immune, inflammatory, and coagulation systems are activated and appear to benefit the patient. Cytokines and inflammatory mediators account for an undefined part of the clinical signs. Rickettsial lipopolysaccharide is biologically relatively nontoxic and does not appear to cause the pathogenic effects of these rickettsial diseases.

The pathologic effects of these rickettsial diseases originate from the multifocal areas of endothelial injury with loss of intravascular fluid into tissue spaces (edema), resultant low blood volume, reduced perfusion of the organs, and disordered function of the tissues with damaged blood vessels (e.g., encephalitis, pneumonitis, and hemorrhagic rash).

DIAGNOSIS

Diagnosis of rickettsial infections is often difficult. The clinical signs and symptoms (e.g., fever, headache, nausea, vomiting, and muscle aches) resemble many other diseases during the early stages when antibiotic treatment is most effective. A history of exposure to the appropriate vector tick, louse, flea, or mite is helpful but cannot be relied upon. Observation of a rash, which usually appears on or after day 3 of illness, should suggest the possibility of a rickettsial infection but, of course, may occur in many other diseases also. Knowledge of the seasonal and geographic epidemiology of rickettsioses is useful, but is inconclusive for the individual patient. Except for epidemic louse-borne typhus, rickettsial diseases strike mostly as isolated single cases in any particular neighbourhood. Therefore, clinico-epidemiologic diagnosis is ultimately a matter of suspicion, empirical treatment, and later laboratory confirmation of the specific diagnosis.

Because rickettsiae are both fastidious and hazardous, few laboratories undertake their isolation and diagnostic identification. Some laboratories are able to identify rickettsiae by immunohistology in skin biopsies as a timely, acute diagnostic procedure, but to establish the diagnosis physicians usually rely on serologic demonstration of the development of antibodies to rickettsial antigens in serum collected after the patient has recovered. Currently, assays that demonstrate antibodies to rickettsial antigens themselves (e.g., the indirect fluorescence antibody test or latex agglutination) are preferable to the nonspecific, insensitive Weil-Felix test that is based on the cross-reactive antigens of OX-19 and OX-2 strains of *Proteus vulgaris*.

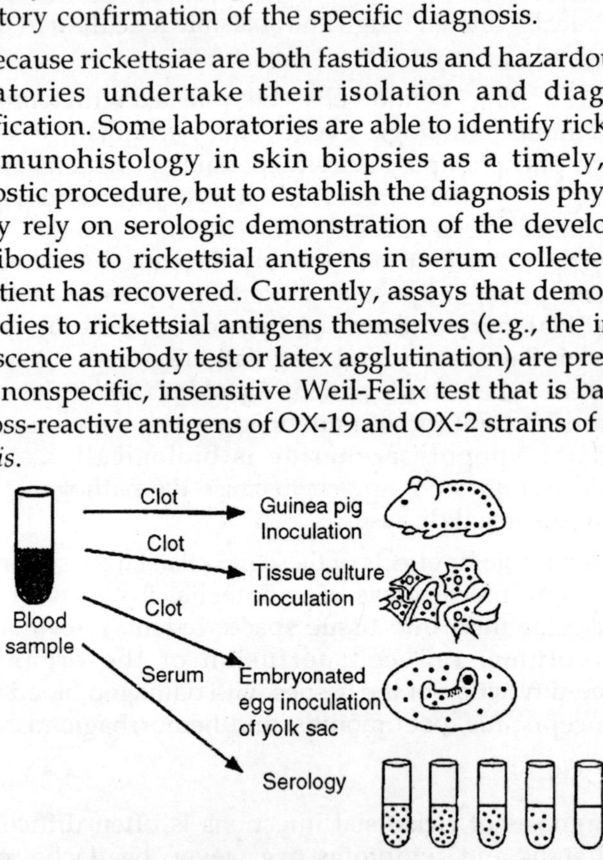

Fig. Laboratory methods used in confirming a diagnosis of rickettsial infection.

Control

Although early treatment with doxycycline, tetracycline, or chloramphenicol is effective in controlling the infection in

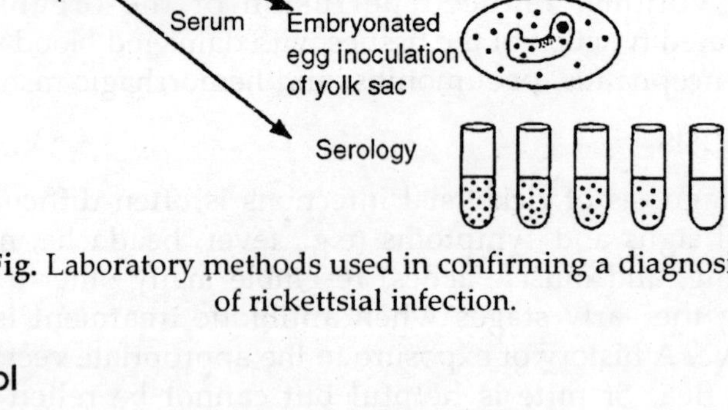

the individual patient, this action has no effect on rickettsiae in their natural ecologic niches. Human infections are prevented by control of the vector and reservoir hosts. Massive delousing with insecticide can abort an epidemic of typhus fever. Prevention of attachment of ticks and their removal before they have injected rickettsiae into the skin reduces the likelihood of a tick-borne spotted fever. Control of rodent populations and of the access of rats and mice to homes and other buildings may reduce human exposure to *R typhi* and *R akari*.

Vaccines against spotted fever and typhus group rickettsiae have been developed empirically by propagation of rickettsiae in ticks, lice, embryonated hen eggs, and cell culture. Vaccines containing killed organisms have provided incomplete protection. A live attenuated vaccine against epidemic typhus has proved successful, but is accompanied by a substantial incidence of side effects, including a mild form of typhus fever in some persons. The presence of strong immunity in convalescent subjects indicates that vaccine development is feasible, but it requires further study of rickettsial antigens and the effective anti-rickettsial immune response. T-lymphocyte-mediated immune mechanisms, including effects of the lymphokines, gamma interferon tumor necrosis factor, and interleukin-1, seem most important.

Orientia (Rickettsia) tsutsugmushi and Scrub Typhus

Although the agents of scrub typhus bear a single taxonomic name, *Orientia* (*Rickettsia*) *tsutsugamushi*, these interrelated organisms are somewhat heterogeneous and differ strikingly from *Rickettsia* species of the spotted fever and typhus groups.

Clinical Manifestations

Patients with scrub typhus often have only fever, headache, and swollen lymph nodes and in some cases myalgia, gastrointestinal complains, or cough beginning 6 to 21 days following exposure to the vector. Fewer than half of the patients have an eschar at the site where the larval mite fed and the classic rash. The mortality varies but averages 7 per cent without anti-rickettsial treatment.

Structure, Classification, and Antigenic Types

Orientia (*Rickettsia*) *tsutsugamushi* is a very labile rickettsia that is particularly difficult to propagate and separate from the host cells in which it grows. In contrast with spotted fever group and typhus group rickettsiae, *O tsutsugamushi* does not seem to possess lipopolysaccharides, peptidoglycan, a slime layer, or other T-independent antigens. The rickettsial cell wall consists of proteins linked by disulfide bonds. Antigenically distinguishable strains represent only part of what seems to be a great antigenic mosaic. Immunity to infection with the homologous strain wanes within a few years; cross-protective immunity to heterologous strains disappears within a few months. The reasons for this lack of long-term immunity are unclear.

Pathogenesis

Orientia (*Rickettsia*) *tsutsugamushi* is injected into the skin during feeding by a larval trombiculid mite (chigger). An eschar often forms at this location. Rickettsiae spread via the bloodstream and damage the microcirculation of the skin (rash), lungs (pneumonitis), brain (encephalitis), and other organs. The generalized enlargement of lymph nodes is unique among rickettsial diseases. *Orientia* (*Rickettsia*) *tsutsugamushi* is phagocytosed by the host cell, escapes from the phagosome into the cytosol, divides by binary fission, and is released from projections of the cell membrane. The pathogenic mechanism of *O tsutsugamushi* is not known.

Epidemiology

Scrub typhus occurs where chiggers infected with virulent rickettsial strains feed upon humans. *Leptotrombidium deliense* and other mites are found particularly in areas where regrowth of scrub vegetation harbors the *Rattus* species that are hosts for the mites. Some of these foci are quite small and have been referred to as mite islands. Because *O tsutsugamushi* is transmitted transovarially from one generation of mites to the next, these dangerous areas tend to persist for as long as the ecologic conditions, including scrub vegetation, persist. Truly one of the neglected diseases, scrub typhus occurs over a vast area, including Japan, China, the Philippines, New Guinea, Indonesia, other islands of the southwest Pacific Ocean, southeastern Asia, northern Australia, India, Sri Lanka, Pakistan, Russia, and Korea. Recognized in western countries mainly because of large numbers of infections of military personnel during World War II and the Vietnam War, scrub typhus perennially affects native populations. Reinfection and undiagnosed infections are highly prevalent. Mortality ranges from 0 to 35 per cent and has not been correlated with any specific factor.

Diagnosis

The cases with fever, headache, eschar, and rash are far outnumbered by cases that lack rash or eschar. Such cases are usually misdiagnosed. Laboratory diagnosis is unavailable in many areas where scrub typhus occurs. Isolation of rickettsiae requires inoculation of mice or cell culture. Serologic diagnosis is made by specific methods (indirect fluorescence antibody test or enzyme immunoassay) or by the older method of demonstrating cross-reactive antibodies that agglutinate the OXK strain of *P mirabilis*.

Control

Scrub typhus can be treated with doxycycline, tetracycline, or chloramphenicol. Chigger repellents may prevent exposure. Prophylaxis with weekly doses of doxycycline during and for 6 weeks after exposure protects against scrub typhus. Attempts to develop a safe, effective vaccine have failed.

Ehrlichia

According to the evolutionary scheme suggested by 16S rRNA sequence homology, ehrlichiae are genetically related to *Rickettsia* species. The genus *Ehrlichia* contains Gram-negative bacteria that reside in a cluster (morula) within membrane-bound cytoplasmic vacuoles of monocytes and macrophages, or polymorphonuclear leukocytes. Ehrlichiae have been implicated as the agents of diseases of horses (*E risticii* and *E equi*), dogs (*E canis, E ewingii* and *E platys*, a platelet pathogen), and other animals. *Ehrlichia sennetsu* causes a human disease in Japan resembling infectious mononucleosis. Ehrlichiae are unusual in their cell wall structure and they can establish persistent infections.

In 1987 the first case of human ehrlichiosis was reported in the United States. A severely ill man with multiorgan system involvement had morula inclusions demonstrated in peripheral blood leukocytes. Subsequently, cases of human monocytic ehrlichiosis have been documented mainly in eastern and southern states between New Jersey and Texas. The infection has varied from severe and sometimes fatal, mimicking Rocky Mountain spotted fever, to oligosymptomatic and asymptomatic forms. A history of tick bite and the seasonal and geographic occurrence correlate with the predominant tick vector, *Amblyomma americanum*. Illness is often accompanied by leukopenia, thrombocytopenia, and damage to the liver. Lesions include perivasculitis in the central nervous system, kidney, heart, and lungs and granulomas in the bone marrow and liver. Clinical diagnosis is difficult. Laboratory diagnosis by indirect fluorescence antibody assay or polymerase chain reaction is not widely available. *Ehrlichia chaffeensis* morulae are difficult to detect in peripheral blood leukocytes.

In 1994 another serious new infectious disease, human granulocytic ehrlichiosis, was reported. Ehrlichiae seen within morulae in neutrophils in smears of peripheral blood were identified as very closely related to *E phagocytophila* (a European tick-transmitted infection of sheep, cattle, goats, and deer) and *E equi*. The causative organism, like other granulocytic ehrlichiae, has never been cultivated. Human granulocytic ehrlichiosis has been associated with the deer tick, *Ixodes scapularis*, and thus is found as far north as Minnesota, Wisconsin, and New England. Laboratory diagnosis is practically achieved by visualizing morulae in neutrophils, as serology and polymerase chain reaction for the agent are presently research procedures. Sometimes fatal, human granulocytic ehrlichiosis, like *E chaffeensis* infection, can be treated effectively with doxycycline.

Coxiella burnetii and Q Fever

Coxiella burnetii is sufficiently different genetically from the other rickettsial agents that it is placed in a separate group. Unlike the other agents, it is very resistant to chemicals and dehydration. Additionally, its transmission to humans is by the aerosol route, although a tick vector is involved in spread of the bacteria among the reservoir animal hosts.

Clinical Manifestations

Q fever is a highly variable disease, ranging from asymptomatic infection to fatal chronic infective endocarditis. Some patients develop an acute febrile disease that is a nonspecific influenza-like illness or an atypical pneumonia. Other patients are diagnosed after identification of granulomas in their liver or bone marrow. The most serious clinical conditions are chronic *C burnetii* infections, which may involve cardiac valves, the central nervous system, and bone.

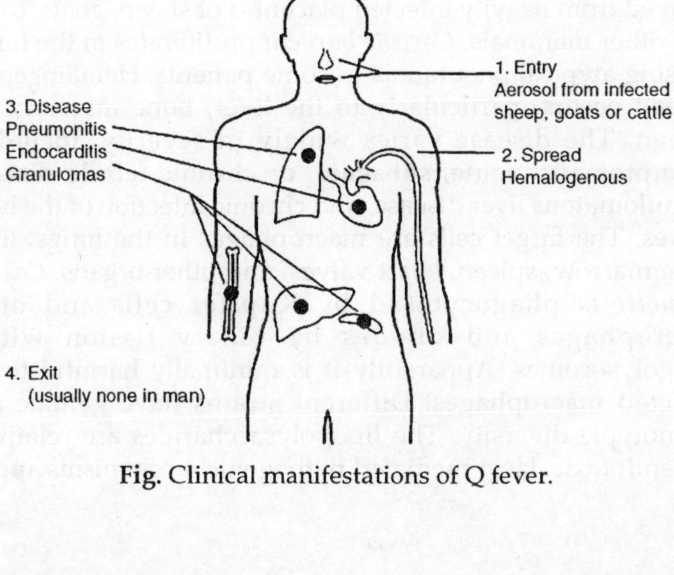

Fig. Clinical manifestations of Q fever.

Structure, Classification, and Antigenic Type

Coxiella burnetii is an obligately intracellular bacterium with some peculiar characteristics. It is small, generally 0.25 µm by 0.5 to 1.25 µm. However, there is considerable ultrastructural pleomorphism, including small- and large-cell variants and possible endospore-like forms, suggesting a hypothetical developmental cycle. Among rickettsiae, *C burnetii* is the most resistant to environmental conditions, is the only species that resides in the phagolysosome, is activated metabolically by low pH, and has a plasmid. The extensive metabolic capacity of *C burnetii* suggests that its obligate intracellular parasitism is a highly evolved state rather than a degenerate condition. The cell wall is typical of Gram-negative bacteria and contains peptidoglycan, proteins, and lipopolysaccharide. When propagated under laboratory conditions in embryonated eggs or cell culture, *C burnetii* undergoes phase variation analogous to the smooth to rough lipopolysaccharide variation of members of the Enterobacteriaceae. Phase I is the form found in nature and in human infections. The phase II variant contains truncated lipopolysaccharide, is avirulent, and is a poor vaccine.

Pathogenesis

Human Q fever follows inhalation of aerosol particles derived from heavily infected placentas of sheep, goats, cattle, and other mammals. *Coxiella burnetii* proliferates in the lungs, causing atypical pneumonia in some patients. Hematogenous spread occurs, particularly to the liver, bone marrow, and spleen. The disease varies widely in severity, including asymptomatic, acute, subacute, or chronic febrile disease, granulomatous liver disease, and chronic infection of the heart valves. The target cells are macrophages in the lungs, liver, bone marrow, spleen, heart valves, and other organs. *Coxiella burnetii* is phagocytosed by Kupffer cells and other macrophages and divides by binary fission within phagolysosomes. Apparently it is minimally harmful to the infected macrophages. Different strains have genetic and phenotypic diversity. The lipopolysaccharides are relatively nonendotoxic. Host-mediated pathogenic mechanisms appear to be important, especially immune and inflammatory reactions, such as T-lymphocyte-mediated granuloma formation.

Epidemiology

Coxiella burnetii infects a wide variety of ticks, domestic livestock, and other wild and domestic mammals and birds throughout the world. Most human infections follow exposure to heavily infected birth products of sheep, goats, and cattle, as occurs on farms, in research laboratories, and in abattoirs. *Coxiella burnetii* is also shed in milk, urine, and feces of infected animals. Animals probably become infected by aerosol and by the bite of any of the 40 species of ticks that carry the organisms.

Diagnosis

Clinical diagnosis depends upon a high index of suspicion, careful evaluation of epidemiologic factors, and ultimately, confirmation by serologic testing. Although *C burnetii* can be isolated by inoculation of animals, embryonated hen eggs, and cell culture, very few laboratories undertake this biohazardous approach. Likewise, the diagnosis is seldom made by visualization of the organisms in infected tissues. Acute Q fever is diagnosed by demonstration of the development of antibodies to protein antigens of *C burnetii* phase II organisms. Chronic Q fever endocarditis is diagnosed by demonstration of a high titer of antibodies, particularly IgG and IgA, against the lipopolysaccharide antigens of *C burnetii* phase I organisms in patients with signs of endocarditis whose routine blood cultures contain no organisms.

Control

Antibiotic treatment is more successful in ameliorating acute, self-limited Q fever than in curing life-threatening chronic endocarditis. Reduction in exposure to these widespread organisms is difficult because some serologically screened animals that have no detectable antibodies to C *burnetii* still shed organisms at parturition. Persons with known occupational hazards (e.g., Australian abattoir workers) have benefitted from a vaccine composed of killed phase I organisms. This vaccine is not readily available, but offers promise for development of safe, effective immunization.

Bartonella

It has been recognized recently that organisms thought to be closely related to rickettsiae such as the louse-borne causative agent of trench fever, *Bartonella* (formerly *Rochalimaea*) *quintana*, in fact, belong in the genus *Bartonella*. These bacteria can be cultivated in cell-free medium and hence do not fit the criterion of definition of rickettsiae as obligately intracellular bacteria. *Bartonella quintana* infections were a serious medical problem during World War I. Soldiers in the trenches were infested with body lice that passed *B quintana* in their feces onto the skin. Individuals who have recovered from trench fever continue to have *R quintana* circulating in this stage of infection and may serve as sources of infection for lice, which can transmit the infection to others.

In association with the AIDS epidemic, another species *B henselae* (in addition to *B quintana)* has been discovered to be the cause of opportunistic infections often masquerading as hemangioma-like lesions of skin and visceral organs, bacillary angiomatosis. *Bartonella henselae* was recognized subsequently to be the long sought after cause of cat scratch disease, which usually manifested as a self-limited enlargement and inflammation of lymph nodes of several months duration in the regional drainage of a cat scratch or bite.

Bartonella bacilliformis transmitted by the sandfly in certain regions of Western South America invades human red blood cells, causing acute, often severe, hemolytic anemia. In chronic infections, there are skin lesions known as *verruga peruana* (Peruvian warts) that are similar to those of bacillary angiomatosis. A Peruvian medical student, Daniel Carrion, proved these lesions to be caused by an infectious agent in 1885 when he fatally inoculated himself with material from a *verruga peruana*. He died of the acute infectious hemolytic anemia known today as Oroya fever or, in his memory, Carrion's disease.

6

Bacteria

A bacterium (plural: bacteria) is a unicellular microorganism. Typically a few micrometres in length, individual bacteria have a wide-range of shapes, ranging from spheres to rods to spirals. Bacteria are ubiquitous in every habitat on Earth, growing in soil, acidic hot springs, radioactive waste, seawater, and deep in the Earth's crust. There are typically 40 million bacterial cells in a gram of soil and a million bacterial cells in a millilitre of fresh water; in all, there are approximately five nonillion (5×10^{30}) bacteria in the world. Bacteria are vital in recycling nutrients, and many important steps in nutrient cycles depend on bacteria, such as the fixation of nitrogen from the atmosphere. However, most of these bacteria have not been characterised, and only about half of the phyla of bacteria have species that can be cultured in the laboratory. The study of bacteria is known as bacteriology, a branch of microbiology.

There are approximately ten times as many bacterial cells as human cells in the human body, with large numbers of bacteria on the skin and in the digestive tract. Although the vast majority of these bacteria are rendered harmless or beneficial by the protective effects of the immune system, a few pathogenic bacteria cause infectious diseases, including cholera, syphilis, anthrax, leprosy and bubonic plague. The most common fatal bacterial diseases are respiratory infections, with tuberculosis alone killing about 2 million people a year, mostly in sub-Saharan Africa. In developed countries, antibiotics are used to treat bacterial infections and in various agricultural processes, so antibiotic resistance is becoming common. In industry, bacteria are important in processes such as wastewater treatment, the production of cheese and yoghurt, and the manufacture of antibiotics and other chemicals.

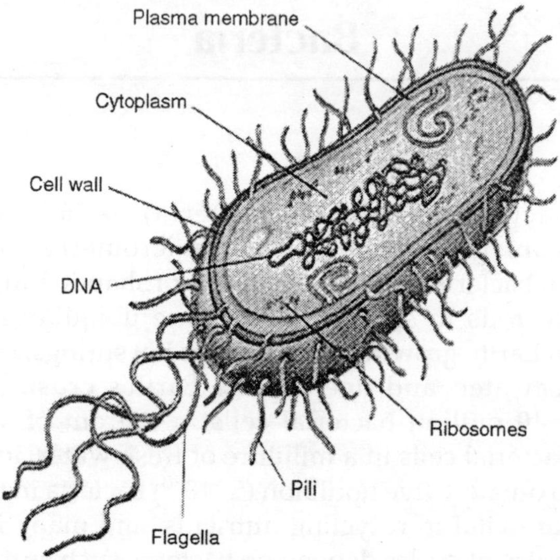

Bacteria are prokaryotes. Unlike cells of animals and other eukaryotes, bacterial cells do not contain a nucleus and rarely harbour membrane-bound organelles. Although the term *bacteria* traditionally included all prokaryotes, the scientific classification changed after the discovery in the 1990s that prokaryotic life consists of two very different groups of organisms that evolved independently from an ancient common ancestor. These evolutionary domains are called Bacteria and Archaea.

HISTORY OF BACTERIOLOGY

The existence of microorganisms was hypothesized during the late Middle Ages. In *The Canon of Medicine* (1020), Abû Alî ibn Sînβ (Avicenna) stated that bodily secretions are contaminated by "foul foreign earthly bodies" before a person becomes infected, but he did not view these bodies as primary causes of disease. When the Black Death bubonic plague reached al-Andalus in the 14th century, Ibn Khatima and Ibn al-Khatib wrote of infectious diseases being caused by contagious entities that enter the human body. These ideas about the contagious nature of some diseases became more popular in Europe during the Renaissance, particularly through the writing of Girolamo Fracastoro.

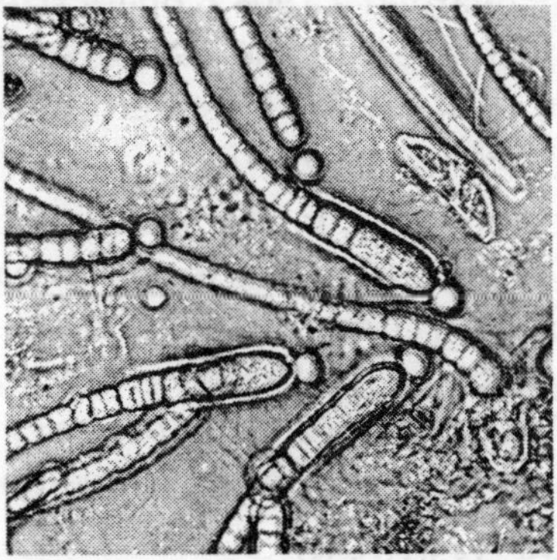

Fig. Microscopic structure of Bacteria

Bacteria were first observed by Anton van Leeuwenhoek in 1676, using a single-lens microscope of his own design. He called them "animalcules" and published his observations in a series of letters to the Royal Society. The name *bacterium* was introduced much later, by Christian Gottfried Ehrenberg in 1828, and is derived from the Greek word βαêôÞñéïí -α, *bacterion -a,* meaning "small staff".

Louis Pasteur demonstrated in 1859 that the fermentation process is caused by the growth of microorganisms, and that this growth is not due to spontaneous generation. (Yeasts and molds, commonly associated with fermentation, are not bacteria, but rather fungi.) Along with his contemporary, Robert Koch, Pasteur was an early advocate of the germ theory of disease. Robert Koch was a pioneer in medical microbiology and worked on cholera, anthrax and tuberculosis. In his research into tuberculosis, Koch finally proved the germ theory, for which he was awarded a Nobel Prize in 1905. In

Koch's postulates, he set out criteria to test if an organism is the cause of a disease; these postulates are still used today.

Though it was known in the nineteenth century that bacteria are the cause of many diseases, no effective antibacterial treatments were available. In 1910, Paul Ehrlich developed the first antibiotic, by changing dyes that selectively stained *Treponema pallidum* – the spirochete that causes syphilis – into compounds that selectively killed the pathogen. Ehrlich had been awarded a 1908 Nobel Prize for his work on immunology, and pioneered the use of stains to detect and identify bacteria, with his work being the basis of the Gram stain and the Ziehl-Neelsen stain.

A major step forward in the study of bacteria was the recognition in 1977 by Carl Woese that archaea have a separate line of evolutionary descent from bacteria. This new phylogenetic taxonomy was based on the sequencing of 16S ribosomal RNA, and divided prokaryotes into two evolutionary domains as part of the three-domain system.

ORIGIN AND EARLY EVOLUTION

The ancestors of modern bacteria were single-celled microorganisms that were the first forms of life to develop on earth, about 4 billion years ago. For about 3 billion years, all organisms were microscopic, and bacteria and archaea were the dominant forms of life. Although bacterial fossils exist, such as stromatolites, their lack of distinctive morphology prevents them from being used to examine the past history of bacterial evolution, or to date the time of origin of a particular bacterial species. However, gene sequences can be used to reconstruct the bacterial phylogeny, and these studies indicate that bacteria diverged first from the archaeal/eukaryotic lineage. The most recent common ancestor of bacteria and archaea was probably a hyperthermophile that lived about 2.5 billion–3.2 billion years ago.

Bacteria were also involved in the second great evolutionary divergence, that of the archaea and eukaryotes. Here, eukaryotes resulted from ancient bacteria entering into endosymbiotic associations with the ancestors of eukaryotic cells, which were themselves possibly related to the Archaea. This involved the engulfment by proto-eukaryotic cells of alpha-proteobacterial symbionts to form either mitochondria or hydrogenosomes, which are still being found in all known Eukarya (sometimes in highly reduced form, e. g. in ancient "amitochondrial" protozoa). Later on, an independent second engulfment by some mitochondria-containing eukaryotes of cyanobacterial-like organisms led to the formation of chloroplasts in algae and plants. There are even some algal groups known that clearly originated from subsequent events of endosymbiosis by heterotrophic eukaryotic hosts engulfing a eukaryotic algae that developed into "second-generation" plastids.

MORPHOLOGY

Bacteria display a wide diversity of shapes and sizes, called *morphologies*. Bacterial cells are about 10 times smaller than eukaryotic cells and are typically 0.5–5.0 micrometres in length. However, a few species–like *Thiomargarita namibiensis* and *Epulopiscium fishelsoni*–are up to half a millimetre long and are visible to the unaided eye. Among the smallest bacteria are members of the genus *Mycoplasma,* which measure only 0.3 micrometres, as small as the largest viruses.

Most bacterial species are either spherical, called cocci (*sing*. coccus, from Greek *kókkos,* grain, seed) or rod-shaped, called bacilli (*sing*. bacillus, from Latin *baculus,* stick). Some rod-shaped bacteria, called vibrio, are slightly curved or comma-shaped; others, can be spiral-shaped, called spirilla, or tightly coiled, called spirochetes. A small number of species even have tetrahedral or cuboidal shapes. This wide variety of shapes is determined by the bacterial cell wall and cytoskeleton, and is important because it can influence the ability of bacteria to acquire nutrients, attach to surfaces, swim through liquids and escape predators.

Many bacterial species exist simply as single cells, others associate in characteristic patterns: *Neisseria* form diploids (pairs), *Streptococcus* form chains, and *Staphylococcus* group together in "bunch of grapes" clusters. Bacteria can also be elongated to form filaments. Filamentous bacteria are often surrounded by a sheath that contains many individual cells; certain types, such as species of the genus *Nocardia*, even form complex, branched filaments, similar in appearance to fungal mycelia.

Bacteria often attach to surfaces and form dense aggregations called biofilms or microbial mats. These films can range from a few micrometers in thickness to up to half a meter in depth, and may contain multiple species of bacteria, protists and archaea. Bacteria living in biofilms display a complex arrangement of cells and extracellular components, forming secondary structures such as microcolonies, through which there are networks of channels to enable better diffusion of nutrients. In natural environments, such as soil or the surfaces of plants, the majority of bacteria are bound to surfaces in biofilms. Biofilms are also important for chronic bacterial infections and infections of implanted medical devices, as bacteria protected within these structures are much harder to kill than individual bacteria.

Even more complex morphological changes are sometimes possible. When starved of amino acids, Myxobacteria detect surrounding cells in a process known as quorum sensing, migrate towards each other, and aggregate to form fruiting bodies up to 500 micrometres long and containing approximately 100,000 bacterial cells. In these fruiting bodies, the bacteria perform separate tasks; this type of cooperation is a simple type of multicellular organisation. About one in 10 cells migrate to the top of these fruiting bodies and differentiate into a specialised dormant state called myxospores, which are more resistant to desiccation and other adverse environmental conditions than are ordinary cells.

CELLULAR STRUCTURE

Intracellular Structures

The bacterial cell is surrounded by a lipid membrane, or cell membrane, which encompasses the contents of the cell and acts as a barrier to hold nutrients, proteins and other essential components of the cytoplasm within the cell. As they are prokaryotes, bacteria do not have membrane-bound organelles in their cytoplasm and thus contain few intracellular structures. They consequently lack a nucleus, mitochondria, chloroplasts and the other organelles present in eukaryotic cells, such as the Golgi apparatus and endoplasmic reticulum. However, recent research is identifying increasing amounts of structural complexity in bacteria, such as the discovery of the prokaryotic cytoskeleton.

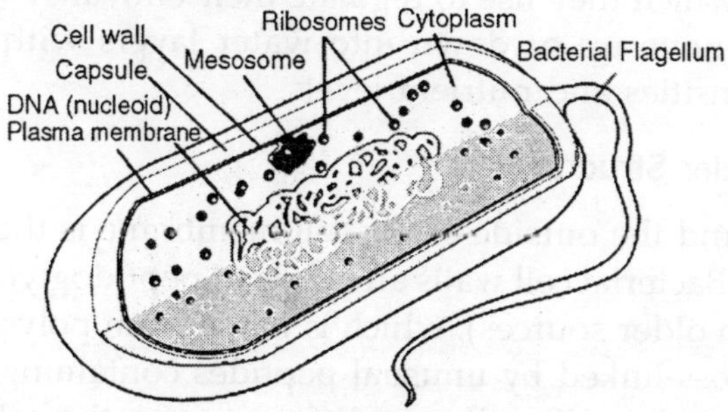

Fig. Cellular structure of a typical bacterial cell

Many important biochemical reactions, such as energy generation, occur due to concentration gradients across membranes, creating a potential difference analogous to a battery. The absence of internal membranes in bacteria means these reactions, such as electron transport, occur across the cell membrane, between the cytoplasm and the periplasmic space. Additionally, while some transporter proteins consume chemical energy, others harness concentration gradients to import nutrients across the cell membrane or to expel undesired molecules from the cytoplasm.

Bacteria do not have a membrane-bound nucleus, and their genetic material is typically a single circular chromosome located in the cytoplasm in an irregularly shaped body called the nucleoid. The nucleoid contains the chromosome with associated proteins and RNA. Like all living organisms, bacteria contain ribosomes for the production of proteins, but the structure of the bacterial ribosome is different from those of eukaryotes and Archaea. The order Planctomycetes are an exception to the general absence of internal membranes in bacteria, because they have a membrane around their nucleoid and contain other membrane-bound cellular structures.

Some bacteria produce intracellular nutrient storage granules, such as glycogen, polyphosphate, sulfur or polyhydroxyalkanoates. These granules enable bacteria to store compounds for later use. Certain bacterial species, such as the photosynthetic Cyanobacteria, produce internal gas vesicles, which they use to regulate their buoyancy - allowing them to move up or down into water layers with different light intensities and nutrient levels.

Extracellular Structures

Around the outside of the cell membrane is the bacterial cell wall. Bacterial cell walls are made of peptidoglycan (called murein in older sources), which is made from polysaccharide chains cross-linked by unusual peptides containing D-amino acids. Bacterial cell walls are different from the cell walls of plants and fungi, which are made of cellulose and chitin, respectively. The cell wall of bacteria is also distinct from that of Archaea, which do not contain peptidoglycan. The cell wall is essential to the survival of many bacteria, and the antibiotic penicillin is able to kill bacteria by inhibiting a step in the synthesis of peptidoglycan.

There are broadly speaking two different types of cell wall in bacteria, called Gram-positive and Gram-negative. The names originate from the reaction of cells to the Gram stain, a test long-employed for the classification of bacterial species.

Gram-positive bacteria possess a thick cell wall containing many layers of peptidoglycan and teichoic acids. In contrast, Gram-negative bacteria have a relatively thin cell wall consisting of a few layers of peptidoglycan surrounded by a second lipid membrane containing lipopolysaccharides and lipoproteins. Most bacteria have the Gram-negative cell wall, and only the Firmicutes and Actinobacteria (previously known as the low G+C and high G+C Gram-positive bacteria, respectively) have the alternative Gram-positive arrangement. These differences in structure can produce differences in antibiotic susceptibility; for instance, vancomycin can kill only Gram-positive bacteria and is ineffective against Gram-negative pathogens, such as *Haemophilus influenzae* or *Pseudomonas aeruginosa*.

In many bacteria an S-layer of rigidly arrayed protein molecules covers the outside of the cell. This layer provides chemical and physical protection for the cell surface and can act as a macromolecular diffusion barrier. S-layers have diverse but mostly poorly understood functions, but are known to act as virulence factors in *Campylobacter* and contain surface enzymes in *Bacillus stearothermophilus*.

'Flagella are rigid protein structures, about 20 nanometres in diameter and up to 20 micrometres in length, that are used for motility. Flagella are driven by the energy released by the transfer of ions down an electrochemical gradient across the cell membrane.

Fimbriae are fine filaments of protein, just 2-10 nanometres in diameter and up to several micrometers in length. They are distributed over the surface of the cell, and resemble fine hairs when seen under the electron microscope. Fimbriae are believed to be involved in attachment to solid surfaces or to other cells and are essential for the virulence of some bacterial pathogens. Pili (*sing.* pilus) are cellular appendages, slightly larger than fimbriae, that can transfer genetic material between bacterial cells in a process called conjugation. .

Capsules or slime layers are produced by many bacteria to surround their cells, and vary in structural complexity: ranging from a disorganised slime layer of extra-cellular polymer, to a highly structured capsule or glycocalyx. These structures can protect cells from engulfment by eukaryotic cells, such as macrophages. They can also act as antigens and be involved in cell recognition, as well as aiding attachment to surfaces and the formation of biofilms.

The assembly of these extracellular structures is dependent on bacterial secretion systems. These transfer proteins from the cytoplasm into the periplasm or into the environment around the cell. Many types of secretion systems are known and these structures are often essential for the virulence of pathogens, so are intensively studied.

ENDOSPORES

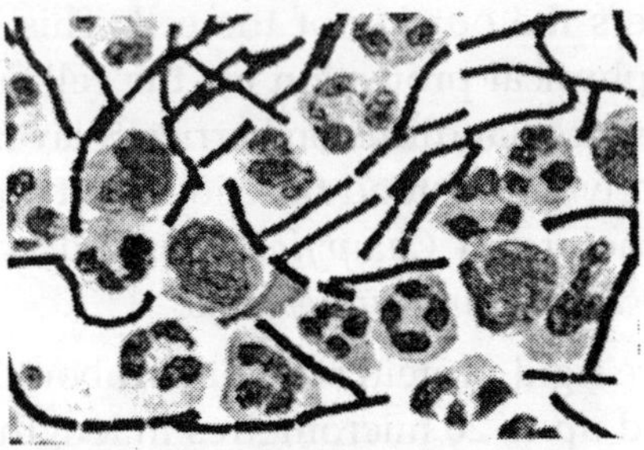

Fig. Bacillus anthracis growing in cerebrospinal fluid.

Certain genera of Gram-positive bacteria, such as *Bacillus*, *Clostridium*, *Sporohalobacter*, *Anaerobacter* and *Heliobacterium*, can form highly resistant, dormant structures called endospores.[70] In almost all cases, one endospore is formed and this is not a reproductive process, although *Anaerobacter* can make up to seven endospores in a single cell. Endospores have a central core of cytoplasm containing DNA and ribosomes surrounded by a cortex layer and protected by an impermeable and rigid coat.

Endospores show no detectable metabolism and can survive extreme physical and chemical stresses, such as high levels of UV light, gamma radiation, detergents, disinfectants, heat, pressure and desiccation. In this dormant state, these organisms may remain viable for millions of years, and endospores even allow bacteria to survive exposure to the vacuum and radiation in space. Endospore-forming bacteria can also cause disease: like, anthrax can be contracted by the inhalation of *Bacillus anthracis* endospores, and contamination of deep puncture wounds with *Clostridium tetani* endospores causes tetanus.

METABOLISM

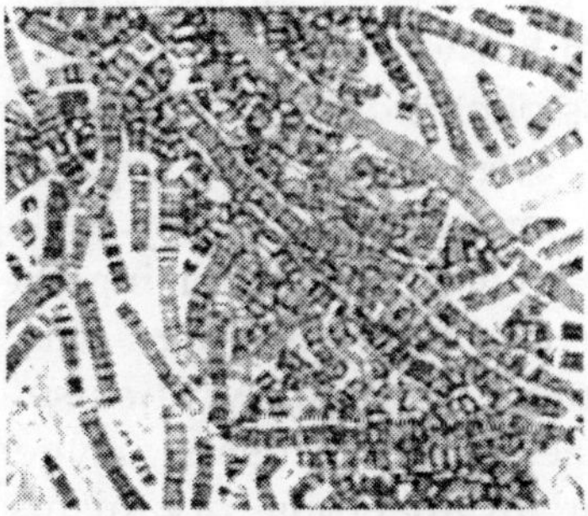

Fig. Filaments of photosynthetic cyanobacteria.

In contrast to higher organisms, bacteria exhibit an extremely wide variety of metabolic types. The distribution of metabolic traits within a group of bacteria has traditionally been used to define their taxonomy, but these traits often do not correspond with modern genetic classifications. Bacterial metabolism is classified on the basis of three major criteria: the kind of energy used for growth, the source of carbon, and the electron donors used for growth. An additional criterion of respiratory microorganisms are the electron acceptors used for aerobic or anaerobic respiration.

Carbon metabolism in bacteria is either heterotrophic, where organic carbon compounds are used as carbon sources, or autotrophic, meaning that cellular carbon is obtained by fixing carbon dioxide. Typical autotrophic bacteria are phototrophic cyanobacteria, green sulfur-bacteria and some purple bacteria, but also many chemolithotrophic species, e. g. nitrifying or sulfur-oxidising bacteria. Energy metabolism of bacteria is either based on phototrophy, the use of light through photosynthesis, or on chemotrophy, the use of chemical substances for energy, which are mostly oxidised at the expense of oxygen or alternative electron acceptors (aerobic/anaerobic respiration).

Finally, bacteria are further divided into lithotrophs that use inorganic electron donors and organotrophs that use organic compounds as electron donors. Chemotrophic organisms use the respective electron donors for energy conservation (by aerobic/anaerobic respiration or fermentation) and biosynthetic reactions (e.g. carbon dioxide fixation), whereas phototrophic organisms use them only for biosynthetic purposes. Respiratory organisms use chemical compounds as a source of energy by taking electrons from the reduced substrate and transferring them to a terminal electron acceptor in a redox reaction. This reaction releases energy that can be used to synthesise ATP and drive metabolism. In aerobic organisms, oxygen is used as the electron acceptor. In anaerobic organisms other inorganic compounds, such as nitrate, sulfate or carbon dioxide are used as electron acceptors. This leads to the ecologically important processes of denitrification, sulfate reduction and acetogenesis, respectively.

Another way of life of chemotrophs in the absence of possible electron acceptors is fermentation, where the electrons taken from the reduced substrates are transferred to oxidised intermediates to generate reduced fermentation products (e. g. lactate, ethanol, hydrogen, butyrate). Fermentation is possible, because the energy content of the substrates is higher than that of the products, which allows the organisms to synthesise ATP and drive their metabolism.

These processes are also important in biological responses to pollution; sulfate-reducing bacteria are largely responsible for the production of the highly toxic forms of mercury (methyl- and dimethylmercury) in the environment. Non-respiratory anaerobes use fermentation to generate energy and reducing power, secreting metabolic by-products (such as ethanol in brewing) as waste. Facultative anaerobes can switch between fermentation and different terminal electron acceptors depending on the environmental conditions in which they find themselves.

Lithotrophic bacteria can use inorganic compounds as a source of energy. Common inorganic electron donors are hydrogen, carbon monoxide, ammonia (leading to nitrification), ferrous iron and other reduced metal ions, and several reduced sulfur compounds. Unusually, the gas methane can be used by methanotrophic bacteria as both a source of electrons and a substrate for carbon anabolism. In both aerobic phototrophy and chemolithotrophy, oxygen is used as a terminal electron acceptor, while under anaerobic conditions inorganic compounds are used instead. Most lithotrophic organisms are autotrophic, whereas organotrophic organisms are heterotrophic.

In addition to fixing carbon dioxide in photosynthesis, some bacteria also fix nitrogen gas (nitrogen fixation) using the enzyme nitrogenase. This environmentally important trait can be found in bacteria of nearly all the metabolic types listed above, but is not universal.

GROWTH AND REPRODUCTION

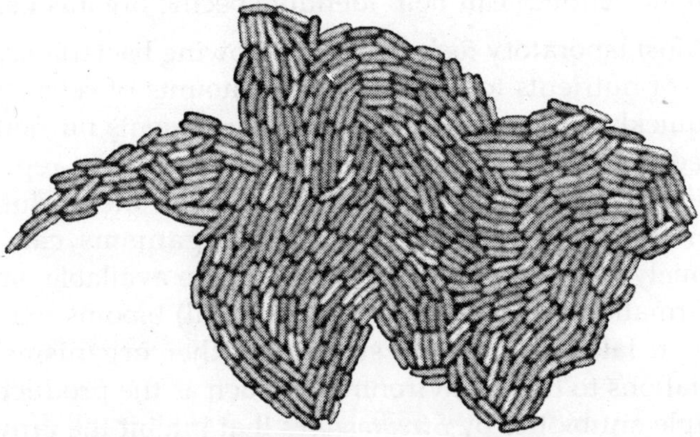

Fig. A growing colony of *Escherichia coli* cells.

Unlike multicellular organisms, increases in the size of bacteria (cell growth) and their reproduction by cell division are tightly linked in unicellular organisms. Bacteria grow to a fixed size and then reproduce through binary fission, a form of asexual reproduction. Under optimal conditions, bacteria can grow and divide extremely rapidly, and bacterial populations can double as quickly as every 9.8 minutes. In cell division, two identical clone daughter cells are produced. Some bacteria, while still reproducing asexually, form more complex reproductive structures that facilitate the dispersal of the newly formed daughter cells. Fruiting body formed by *Myxobacteria* and arial hyphae formation by *Streptomyces*, or budding. Budding involves a cell forming a protrusion that breaks away and produces a daughter cell.

In the laboratory, bacteria are usually grown using solid or liquid media. Solid growth media such as agar plates are used to isolate pure cultures of a bacterial strain. However, liquid growth media are used when measurement of growth or large volumes of cells are required. Growth in stirred liquid media occurs as an even cell suspension, making the cultures easy to divide and transfer, although isolating single bacteria from liquid media is difficult. The use of selective media (media with specific nutrients added or deficient, or with antibiotics added) can help identify specific organisms.

Most laboratory techniques for growing bacteria use high levels of nutrients to produce large amounts of cells cheaply and quickly. However, in natural environments nutrients are limited, meaning that bacteria cannot continue to reproduce indefinitely. This nutrient limitation has led the evolution of different growth strategies. Some organisms can grow extremely rapidly when nutrients become available, such as the formation of algal (and cyanobacterial) blooms that often occur in lakes during the summer. Other organisms have adaptations to harsh environments, such as the production of multiple antibiotics by *Streptomyces* that inhibit the growth of competing microorganisms. In nature, many organisms live in communities (e.g. biofilms) which may allow for increased supply of nutrients and protection from environmental stresses. These relationships can be essential for growth of a particular organism or group of organisms (syntrophy).

Bacterial growth follows three phases. When a population of bacteria first enter a high-nutrient environment that allows growth, the cells need to adapt to their new environment. The first phase of growth is the lag phase, a period of slow growth when the cells are adapting to the high-nutrient environment and preparing for fast growth. The lag phase has high biosynthesis rates, as growth factors(enzymes) are produced. The second phase of growth is the logarithmic phase (log phase), also known as the exponential phase. The log phase is marked by rapid exponential growth. The rate at which cells grow during this phase is known as the *growth rate* (*k*), and the time it takes the cells to double is known as the *generation time* (*g*). During log phase, nutrients are metabolised at maximum speed until one of the nutrients is depleted and starts limiting growth. The final phase of growth is the stationary phase and is caused by depleted nutrients. The cells reduce their metabolic activity and consume non-essential cellular proteins. The stationary phase is a transition from rapid growth to a stress response state and there is increased expression of genes involved in DNA repair, antioxidant metabolism and nutrient transport.

GENETICS

Most bacteria have a single circular chromosome that can range in size from only 160,000 base pairs in the endosymbiotic bacteria *Candidatus Carsonella ruddii*, to 12,200,000 base pairs in the soil-dwelling bacteria *Sorangium cellulosum*. Spirochaetes of the genus *Borrelia* are a notable exception to this arrangement, with bacteria such as *Borrelia burgdorferi*, the cause of Lyme disease, containing a single linear chromosome. Bacteria may also contain plasmids, which are small extra-chromosomal DNAs that may contain genes for antibiotic resistance or virulence factors. Another type of bacterial DNA are integrated viruses (bacteriophages). Many types of bacteriophage exist, some simply infect and lyse their host bacteria, while others insert into the bacterial chromosome. A bacteriophage can contain genes that contribute to its host's phenotype: In the evolution of *Escherichia coli* O157:H7 and *Clostridium botulinum*, the toxin genes in an integrated phage converted a harmless ancestral bacteria into a lethal pathogen.

Bacteria, as asexual organisms, inherit identical copies of their parent's genes (i.e., they are clonal). However, all bacteria can evolve by selection on changes to their genetic material DNA caused by genetic recombination or mutations.

Mutations come from errors made during the replication of DNA or from exposure to mutagens. Mutation rates vary widely among different species of bacteria and even among different clones of a single species of bacteria. Genetic changes in bacterial genomes come from either random mutation during replication or "stress-directed mutation", where genes involved in a particular growth-limiting process have an increased mutation rate.

Some bacteria also transfer genetic material between cells. This can occur in three main ways. Firstly, bacteria can take up exogenous DNA from their environment, in a process called transformation. Genes can also be transferred by the process of transduction, when the integration of a bacteriophage introduces foreign DNA into the chromosome. The third method of gene transfer is bacterial conjugation, where DNA is transferred through direct cell contact. This gene acquisition from other bacteria or the environment is called horizontal gene transfer and may be common under natural conditions. Gene transfer is particularly important in antibiotic resistance as it allows the rapid transfer of resistance genes between different pathogens.

MOVEMENT

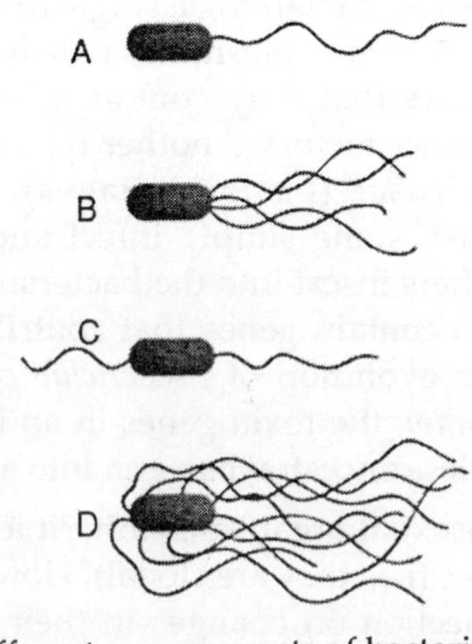

Fig. The different arrangements of bacterial flagella: A-Monotrichous; B-Lophotrichous; C-Amphitrichous; D-Peritrichous;

Motile bacteria can move using flagella, bacterial gliding, twitching motility or changes of buoyancy. In twitching motility, bacterial use their type IV pili as a grappling hook, repeatedly extending it, anchoring it and then retracting it with remarkable force (>80 pN).

Bacterial species differ in the number and arrangement of flagella on their surface; some have a single flagellum (monotrichous), a flagellum at each end (amphitrichous), clusters of flagella at the poles of the cell (lophotrichous), while others have flagella distributed over the entire surface of the cell (peritrichous). The bacterial flagella is the best-understood motility structure in any organism and is made of about 20 proteins, with approximately another 30 proteins required for its regulation and assembly. The flagellum is a rotating structure driven by a motor at the base that uses the electrochemical gradient across the membrane for power. This motor drives the motion of the filament, which acts as a propeller. Many bacteria have two distinct modes of movement: forward movement (swimming) and tumbling. The tumbling allows them to reorient and makes their movement a three-dimensional random walk. The flagella of a unique group of bacteria, the spirochaetes, are found between two membranes in the periplasmic space. They have a distinctive helical body that twists about as it moves.

Motile bacteria are attracted or repelled by certain stimuli in behaviours called *taxes*: these include chemotaxis, phototaxis and magnetotaxis. In one peculiar group, the myxobacteria, individual bacteria move together to form waves of cells that then differentiate to form fruiting bodies containing spores. The myxobacteria move only when on solid surfaces, unlike *E. coli* which is motile in liquid or solid media.

Several *Listeria* and *Shigella* species move inside host cells by usurping the cytoskeleton, which is normally used to move organelles inside the cell. By promoting actin polymerization at one pole of their cells, they can form a kind of tail that pushes them through the host cell's cytoplasm.

CLASSIFICATION AND IDENTIFICATION

Classification seeks to describe the diversity of bacterial species by naming and grouping organisms based on similarities. Bacteria can be classified on the basis of cell structure, cellular metabolism or on differences in cell components such as DNA, fatty acids, pigments, antigens and quinones. While these schemes allowed the identification and classification of bacterial strains, it was unclear whether these differences represented variation between distinct species or between strains of the same species. This uncertainty was due to the lack of distinctive structures in most bacteria, as well as lateral gene transfer between unrelated species. Due to lateral gene transfer, some closely related bacteria can have very different morphologies and metabolisms. To overcome this uncertainty, modern bacterial classification emphasizes molecular systematics, using genetic techniques such as guanine cytosine ratio determination, genome-genome hybridization, as well as sequencing genes that have not undergone extensive lateral gene transfer, such as the rRNA gene. Classification of bacteria is determined by publication in the International Journal of Systematic Bacteriology, and Bergey's Manual of Systematic Bacteriology.

The term "bacteria" was traditionally applied to all microscopic, single-celled prokaryotes. However, molecular systematics showed prokaryotic life to consist of two separate domains, originally called *Eubacteria* and *Archaebacteria,* but now called *Bacteria* and *Archaea* that evolved independently from an ancient common ancestor. These two domains, along with Eukarya, are the basis of the three-domain system, which is currently the most widely used classification system in bacteriology. However, due to the relatively recent introduction of molecular systematics and the analysis of genome sequences, bacterial classification remains a changing and expanding field. A few biologists argue that Archaea evolved from Gram-positive bacteria.

Identification of bacteria in the laboratory is particularly relevant in medicine, where the correct treatment is determined by the bacterial species causing an infection. Consequently, the need to identify human pathogens was a major impetus for the development of techniques to identify bacteria.

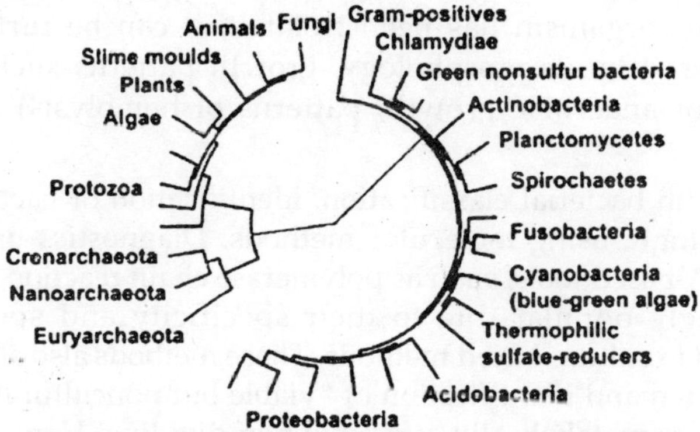

Fig. Phylogenetic tree showing the incredible diversity of bacteria, compared to other organisms.

The Gram stain, developed in 1884 by Hans Christian Gram, characterises bacteria based on the structural characteristics of their cell walls. The thick layers of peptidoglycan in the "Gram-positive" cell wall stain purple, while the thin "Gram-negative" cell wall appears pink. By combining morphology and Gram-staining, most bacteria can be classified as belonging to one of four groups (Gram-positive cocci, Gram-positive bacilli, Gram-negative cocci and Gram-negative bacilli). Some organisms are best identified by stains other than the Gram stain, particularly mycobacteria or *Nocardia*, which show acid-fastness on Ziehl–Neelsen or similar stains. Other organisms may need to be identified by their growth in special media, or by other techniques, such as serology.

Culture techniques are designed to promote the growth and identify particular bacteria, while restricting the growth of the other bacteria in the sample. Often these techniques are designed for specific specimens; a sputum sample will be treated to identify organisms that cause pneumonia, while stool specimens are cultured on selective media to identify organisms that cause diarrhoea, while preventing growth of non-pathogenic bacteria. Specimens that are normally sterile, such as blood, urine or spinal fluid, are cultured under conditions designed to grow all possible organisms. Once a pathogenic organism has been isolated, it can be further characterised by its morphology, growth patterns such as (aerobic or anaerobic growth, patterns of hemolysis) and staining.

As with bacterial classification, identification of bacteria is increasingly using molecular methods. Diagnostics using such DNA-based tools, such as polymerase chain reaction, are increasingly popular due to their specificity and speed, compared to culture-based methods. These methods also allow the detection and identification of "viable but nonculturable" cells that are metabolically active but non-dividing. However, even using these improved methods, the total number of bacterial species is not known and cannot even be estimated with any certainty. Attempts to quantify bacterial diversity have ranged from 10^7 to 10^9 total species, but even these diverse estimates may be out by many orders of magnitude.

INTERACTIONS WITH OTHER ORGANISMS

Despite their apparent simplicity, bacteria can form complex associations with other organisms. These symbiotic associations can be divided into parasitism, mutualism and commensalism. Due to their small size, commensal bacteria are ubiquitous and grow on animals and plants exactly as they will grow on any other surface. However, their growth can be increased by warmth and sweat, and large populations of these organisms in humans are the cause of body odor.

Mutualists

Certain bacteria form close spatial associations that are essential for their survival. One such mutualistic association, called interspecies hydrogen transfer, occurs between clusters of anaerobic bacteria that consume organic acids such as butyric acid or propionic acid and produce hydrogen, and methanogenic Archaea that consume hydrogen. The bacteria in this association are unable to consume the organic acids as this reaction produces hydrogen that accumulates in their surroundings. Only the intimate association with the hydrogen-consuming Archaea keeps the hydrogen concentration low enough to allow the bacteria to grow.

In soil, microorganisms which reside in the rhizosphere (a zone that includes the root surface and the soil that adheres to the root after gentle shaking) carry out nitrogen fixation, converting nitrogen gas to nitrogenous compounds. This serves to provide an easily absorbable form of nitrogen for many plants, which cannot fix nitrogen themselves. Many other bacteria are found as symbionts in humans and other organisms. The presence of over 1,000 bacterial species in the normal human gut flora of the intestines can contribute to gut immunity, synthesise vitamins such as folic acid, vitamin K and biotin, convert milk protein to lactic acid, as well as fermenting complex undigestible carbohydrates. The presence of this gut flora also inhibits the growth of potentially pathogenic bacteria (usually through competitive exclusion) and these beneficial bacteria are consequently sold as probiotic dietary supplements.

Pathogens

If bacteria form a parasitic association with other organisms, they are classed as pathogens. Pathogenic bacteria are a major cause of human death and disease and cause infections such as tetanus, typhoid fever, diphtheria, syphilis, cholera, food-borne illness, leprosy and tuberculosis. A pathogenic cause for a known medical disease may only be discovered many years after, as was the case with Helicobacter pylori and peptic ulcer disease. Bacterial diseases are also important in agriculture, with bacteria causing leaf spot, fireblight and wilts in plants, as well as Johne's disease, mastitis, salmonella and anthrax in farm animals.

Each species of pathogen has a characteristic spectrum of interactions with its human hosts. Some organisms, such as *Staphylococcus* or *Streptococcus*, can cause skin infections, pneumonia, meningitis and even overwhelming sepsis, a systemic inflammatory response producing shock, massive vasodilation and death. Yet these organisms are also part of the normal human flora and usually exist on the skin or in the nose without causing any disease at all. Other organisms invariably cause disease in humans, such as the Rickettsia, which are obligate intracellular parasites able to grow and reproduce only within the cells of other organisms. One species of Rickettsia causes typhus, while another causes Rocky Mountain spotted fever. Chlamydia, another phylum of obligate intracellular parasites, contains species that can cause pneumonia, or urinary tract infection and may be involved in coronary heart disease. Finally, some species such as *Pseudomonas aeruginosa*, *Burkholderia cenocepacia*, and *Mycobacterium avium* are opportunistic pathogens and cause disease mainly in people suffering from immunosuppression or cystic fibrosis.

Bacterial infections may be treated with antibiotics, which are classified as bacteriocidal if they kill bacteria, or bacteriostatic if they just prevent bacterial growth. There are many types of antibiotics and each class inhibits a process that is different in the pathogen from that found in the host. Antibiotics produce selective toxicity is chloramphenicol and puromycin, which inhibit the bacterial ribosome, but not the structurally different eukaryotic ribosome. Antibiotics are used both in treating human disease and in intensive farming to promote animal growth, where they may be contributing to the rapid development of antibiotic resistance in bacterial populations. Infections can be prevented by antiseptic measures such as sterilizating the skin prior to piercing it with the needle of a syringe, and by proper care of indwelling catheters. Surgical and dental instruments are also sterilized to prevent contamination and infection by bacteria. Disinfectants such as bleach are used to kill bacteria or other pathogens on surfaces to prevent contamination and further reduce the risk of infection.

SIGNIFICANCE IN TECHNOLOGY AND INDUSTRY

Bacteria, often *Lactobacillus* in combination with yeasts and molds, have been used for thousands of years in the preparation of fermented foods such as cheese, pickles, soy sauce, sauerkraut, vinegar, wine and yoghurt.

The ability of bacteria to degrade a variety of organic compounds is remarkable and has been used in waste processing and bioremediation. Bacteria capable of digesting the hydrocarbons in petroleum are often used to clean up oil spills. Fertilizer was added to some of the beaches in Prince William Sound in an attempt to promote the growth of these naturally occurring bacteria after the infamous 1989 *Exxon Valdez* oil spill. These efforts were effective on beaches that were not too thickly covered in oil. Bacteria are also used for the bioremediation of industrial toxic wastes. In the chemical industry, bacteria are most important in the production of enantiomerically pure chemicals for use as pharmaceuticals or agrichemicals.

Bacteria can also be used in the place of pesticides in the biological pest control. This commonly involves *Bacillus thuringiensis* (also called BT), a Gram-positive, soil dwelling bacterium. Subspecies of this bacteria are used as a Lepidopteran-specific insecticides under trade names such as Dipel and Thuricide. Because of their specificity, these pesticides are regarded as environmentally friendly, with little or no effect on humans, wildlife, pollinators and most other beneficial insects.

Because of their ability to quickly grow and the relative ease with which they can be manipulated, bacteria are the workhorses for the fields of molecular biology, genetics and biochemistry. By making mutations in bacterial DNA and examining the resulting phenotypes, scientists can determine the function of genes, enzymes and metabolic pathways in bacteria, then apply this knowledge to more complex organisms. This aim of understanding the biochemistry of a cell reaches its most complex expression in the synthesis of huge amounts of enzyme kinetic and gene expression data into mathematical models of entire organisms. This is achievable in some well-studied bacteria, with models of *Escherichia coli* metabolism now being produced and tested.

This understanding of bacterial metabolism and genetics allows the use of biotechnology to bioengineer bacteria for the production of therapeutic proteins, such as insulin, growth factors, or antibodies.

INTERSETIN FACT REGARDING BACTERIA

Bacteria Eat Pollution, Generate Electricity

Scientists have long studied bacteria that can clean up toxic waste by eating it. Other bacteria have been employed to produce electricity. Now scientists have found a two-for-one deal in bacteria that will eat toxic chemicals 24/7 and make electricity to boot. "The bacteria are capable of continuously generating electricity at levels that could be used to operate small electronic devices," says Charles Milliken of the Medical University of South Carolina, who conducted the research with colleague Harold May. "As long as the bacteria are fed fuel they are able to produce electricity 24 hours a day."

The findings were presented at the General Meeting of the American Society for Microbiology. The new study involved Desulfitobacteria, already known for their ability to breakdown and detoxifies some of the most problematic environmental pollutants, including PCBs and some chemical solvents. "These bacteria are very diverse in their metabolic capabilities, including the food that they can consume. This means that these bacteria can convert a large number of different food sources into electricity," says Milliken. "The technology could be used to assist in the reclamation of wastewaters, thereby resulting in the removal of waste and generation of electricity."

The bacteria perform their useful tasks while in spore form, a dormant stage of growth that can handle extreme heat, radiation and lack of water – all useful traits for an organism that might be employed in some of the worst manmade environments.

Bacteria in Yogurt Modified to Fight HIV Infection

Bacteria in yogurt have been modified to deliver a drug that blocks HIV infection. In their natural state, the bacteria

(*Lactococcus lactis*) produce lactic acid and are used to make cheese and yogurt. It is not harmful to humans. Research based at Brown Medical School in Rhode Island altered the genetics of the bacteria so they generate cyanovirin, a drug that has prevented HIV infection in monkeys and human cells. Cyanovirin binds to sugar molecules that are attached to the HIV virus, blocking a receptor used by HIV to infect cells. "It's basically passive immunisation," said Sean Hanniffy, a member of the research team from the Institute of Food Research in Norwich, UK. Cyanovirin could be put into gels that women would apply to the vagina before sex. Or, an oral dose might provide long-term protection, the researchers speculate. They note that there could be public resistance to the whole idea of genetic modification.

Microbe and Machine Merged to Create First 'Cellborg'

Fully merging microbe and machine for the first time, scientists have created gold-plated bacteria that can sense humidity. The breakthrough is the first "cellborg" in what might become an array of devices that could sense dangerous gases or other hazardous substances. The bioelectronic device swells and contracts in response to how much water vapour is in the air. It's called a cellborg humidity sensor, and it is at least four times more sensitive than those that are solely electronic. It even works even when its biological parts are long dead.

How it was Made

Scientists first coated a silicon chip with a layer of live *Bacillus cereus* bacteria. Some of the long, rod-shaped microbes lodged between two etched electrodes on the chip's surface, forming a bridge. The chip was then washed in a solution containing tiny gold particles, each one about 30 nanometers across. A nanometer is one billionth of a meter. A human hair is roughly 100,000 nanometers wide. The gold nanoparticles attached to long hair-like proteins on the surface of the bacteria, transforming them into gold-plated bridges that completed an electronic circuit.

The hair-like proteins are called teichoic acid molecules. They are negatively charged and provide a surface for the positively-charged gold nanoparticles to attach to. Without them, the gold nanoparticles would repel one another due to their like-charges and no bridge between the two electrodes could ever form. By wrapping themselves around the gold nanoparticles, the teichoic acid molecules therefore act as metal insulators, creating what engineers call a "dielectric barrier." "To any electronic person, that's a field day," a University of Nebraska chemical engineer who led the discovery.

First of its kind

The bodies of the gold-plated bacteria swell as humidity increases and they absorb moisture; they contract when humidity decreases. The swelling causes the gold nanoparticles on the bacteria's surface to grow farther apart, like stickers on an inflating balloon. Even a tiny separation of 0.2 nanometers between the gold nanoparticles was enough to interfere with the flow of electric current between the circuit's two electrodes. That's because the farther apart the gold particles on the bacteria's surface, the harder it becomes for electrons to "hop" between particles and get from one electrode to the other.

The cellborg sensor is extremely sensitive: a drop from 20 per cent to zero humidity results in a 40-fold decrease in current flow. In humidity sensors that are solely electronic, the decrease is only 10-fold. Their hybrid sensor is the first to incorporate microorganisms into an electronic device. In the past, researchers have programmed bacteria to behave like biological computers or created electronic circuits that respond to glowing bacteria as a way to detect chemicals, but in those cases, the line separating microbe and machine was still distinct.

The nearest other attempt to merge the two occurred in March, when researchers at the University of Wisconsin-Madison reported using electrodes to trap and examine bacteria. It might be possible to attach microscopic gold particles to the shell of the bacteria to form "nanoscale gold wire."

Bacteria Zombies

Once assimilated, the gilded bacteria can survive for only about two days, but even when dead, their bodies still swell and contract in response to changes in humidity. They can go on working this way for months. If scientists could coat bacteria with gold nanoparticles without killing them, it might be possible to make cellborg sensors that could *power* an electronic circuit instead of just completing one. Another possibility may be to tweak the bacteria so they respond to things other than humidity. They could be made to swell or contract, when they feed on certain gases or hazardous chemicals.

Bacteria Uses Sonar-Like Strategy to Probe Environment

An infection-causing bacteria uses a unique sonar-like strategy to scan its environment, similar to that used by bats when hunting in the dark. The research solves a 70-year-old medical mystery and could lead to the development of new drugs that target bacterial infections without spurring antibiotic resistance. Bacteria are microscopic single celled organisms, some of which are responsible for human infections and disease. The bacteria *Enterococcus faecalis* has interested researchers ever since 1934, when it was discovered that it could produce a substance called cytolysin that can poison or kill a broad range of organisms.

Scientist have since learned that the toxin is composed of two protein subunits, one small and one large, that are constantly being produced by the bacteria at low levels. It was observed that *E. faecalis* was somehow able to sense the presence of nearby target cells and ramp up toxin production in response, but until now, scientists were in the dark as to how this was accomplished. The research team, led by Michael Gilmore, director of the Schepens Eye Research Institute at Harvard Medical School, solved the mystery when they discovered that the cytolysin subunits also act as a two-part environmental probe. Like armed sentries sent out in pairs, they scout the terrain, and when they come across an enemy, one will attack, while the smaller of the two rushes back with a message for reinforcements.

Here's how it works: If a target cell is present, the larger subunit will bind to it, leaving its smaller half unattended and free to report back to the bacterium. The return signal kicks toxin production from its normally low levels into high gear. In the absence of a target cell, the large subunits latch onto the smaller ones and keep them below the level needed to trigger an increase in toxin production. Although the bacteria do not emit high pitched sounds the way bats do when they navigate in the dark, the basic principle is the same.

The study was detailed in a recent issue of the journal *Science*.

The ability to produce cytolysin probably evolved as a kind of lethal alarm system to help the bacteria establish islands of safety within their local environments and to keep out unwanted visitors, including other bacteria, scientists say. Over time, the adaptation may have evolved to become a useful weapon, giving the bacteria a means of obtaining nutrients – found within other cells, and released only by rupturing them – that would not otherwise be accessible. The discovery could lead to the development of novel "toxin-inhibitor" drugs that could limit the severity of infections. Because the bacteria would not be killed directly, there is less chance of them developing a resistance to the drugs. Such a development would be welcome news in hospitals, as more strains of bacteria are becoming increasingly resistant to even the most advanced antibiotics.

Is Antibacterial Soap Really Necessary?

Some folks at the Harvard Medical School wondered the same things, so they did a review of all the studies on the topic. The more expensive products will indeed clean your hands if you use them right, but "plain old soap and water is still a good way to clean your hands," the researchers write in this month's issue of the *Harvard Health Letter*. In a review of previous studies, they conclude that "antibacterial soaps available to consumers don't add much to hand hygiene," and "antibacterial soaps aren't the all-purpose germ fighters that many people think they are."

The tricks, regardless of which soap you use, is washing for 15 seconds. Doing this with regular soap reduces bacterial counts by about 90 per cent. A second tip: Dry your hands, lest the remaining bacteria spread more easily.

And the alcohol-based hand sanitizers that require no water or towel? "Alcohol doesn't kill everything" and the rubs need to come into contact with all surfaces of your hands, so "using small amounts is really no better than washing with plain soap and water."

NANOBACTERIA

Summary: Nannobacteria are very small living creatures in the 0.05 to 0.2 micrometer range. They are enormously abundant in minerals and rocks, and probably run most of the earth's surface chemistry. Although they form most of the world's biomass, they remain "biota incognita" to the biological world as their genetic relationships, metabolism, and other characteristics remain to be investigated.

Introduction

Nannobacteria are dwarf forms of bacteria, mostly 0.05 to 0.2 micrometers, about one-tenth the diameter and 1/1000 the volume of ordinary bacteria. The word was first published as "nanobacteria" by Richard Y. Morita in 1988, but here the spelling used is "nanno-" to conform with geological usage, e.g., "nannoplankton." In the microbiological world, the minute forms are commonly called "ultramicrobacteria," regarded as stressed or resting forms of big bacteria, and thought to be both rare and dormant. However in soils, sediments, minerals and rocks they are enormously abundant and may perhaps represent a novel life form intermediate between normal bacteria and viruses (which are around 0.01 to 0.02 micrometers). It is contended that they are responsible for a great deal of mineral precipitation as well as conversion of igneous minerals to soils, and the corrosion of metals.

Discovery

The important role of nannobacteria in the mineralogical world was discovered through dumb luck, idle curiosity and random reading. There was no Lifetime Research Plan or This can get me Lots of National Funding idea involved. In the course of this research it was discovered by chance that "normal-sized" bacteria, mainly sulfur-oxidizers, had played a very substantial role in precipitating this stone from the warm springs at Tivoli. A new electron microscope with magnifications up to 100,000X began to reveal hordes of tiny bumps and balls. At first they are passed off as artifacts of sample preparation or laboratory contamination, as had every other scientist who had studied minerals and rocks with the scanning electron microscope (SEM). (One must be very careful to use only a 30-second gold coat on the specimen, or one can indeed produce nannobody-looking artifacts). After a year of doubts, a little reading in Microbiology unearthed the fact that very small cells called "ultramicrobacteria" did in fact exist. With further SEM work, slowly the realization dawned that there really were entombed in minerals enormously abundant cells of this minute size, and in some examples the minerals seemed to be entirely made up of nannobacteria as closely packed as beans in a bag. Sometimes within a single crystal of mineral, part of the crystal would be crowded with nannobacteria and parts would be deserted, belying the idea of artifacts or "that's the way minerals naturally dissolve." Their occurrence in chains and grape-like clusters further attested to their true living status. For several years the research went forward without any assistance or external funding, much to the consternation and puzzled frowns of geological colleagues. First oral presentation of the idea elicited mostly a stony silence.

Calcite from Bullicame hot springs, Viterbo, Italy; etched in dilute HCl. One elliptical "normal-sized" bacterial body is seen, about 1.5 micrometers long. Another normal-sized bacterium at right is only half emerged from its bed of calcite. The hordes of 0.1 micrometer spheres are nannobacteria; in the original SEM photo, the crystal appears to be almost wholly composed of nannobacteria. A small chain of about eight nannobacterial cells appears in the form of a homunculus in the centre. A few spheroidal bodies, 0.2 to 0.6 micrometers, appear to bridge the size gap between ordinary bacteria and nannobacteria.

The idea of the existence of nannobacteria has been greeted with howls of disbelief by the majority of the biological community, who contend that these minute bodies cannot be bacteria because they are too small to contain the necessary genetic machinery for life. Even if they are not "normal" bacteria, they can easily be cultured and those grown for a few days look exactly like those occurring in rocks and minerals as old as 2 billion years or as recent as today, and are dead ringers for those occurring on Mars as to size, shape, and surface features. Although DNA analysis of *mineralised* nannobacteria has yet to be done, some attempt has been made by medical researchers who find "nanobacterial" cells the same size as observed, with cell walls that are very tough and that are resistant to acids, stains and poisons. Because of the tough walls special methods are required to isolate the DNA which occurs as very short strands.

A stub of metallic aluminum is placed in Austin tap water for seven days. Gelatinous clots formed on the stub, and when viewed with the SEM these showed many diverse kinds of nannobacteria; the most noteworthy are the caterpillar-like bodies, about 0.05 micrometers wide, plus single bodies of the same size. These nannobacteria are perhaps metabolizing Al, and are good replicas of the Martian meteorite nannobacteria. Scale bar, 0.5 micrometers.

Nannobacteria lying on top of a flake of chlorite clay; lava flow from Mt. Etna at Acireale, Sicily. These nannobacteria are involved in the precipitation of silicates. Compare these shapes with the nannobacteria grown on the Al stub. Single isolated nannobacteria about 0.03 micrometers, as well as the chains of partially merged bodies. Inorganic minerals do not crystallize like this. Scale bar, 0.5 micrometers.

Occurrence

At the initial discovery site, the hot springs of Viterbo, Lazio, Italy, some nannobacteria are found in untreated samples along with rare bacteria of normal size. However, upon slight etching with HCl, hordes of nannobacteria are revealed entombed in the calcite and aragonite crystals, like peanuts in peanut brittle. They occur in dense clumps and initiate precipitation; once the seed crystal has been formed, the remainder of the crystal continues to form abiotically. The nannobacteria act as catalysts, attracting cations to their negatively charged cell walls, leading to oversaturation on a submicrometer scale and triggering precipitation of $CaCO_3$.

Once they were discovered in the travertines of Viterbo, nannobacteria were soon found in limestones and dolomites ($CaMg(CO_3)_2$) in rocks of all ages back to two billion years old. In some limestones they are abundant, in others they are scarce. Modern oolites of the Bahamas are rich in nannobacterially precipitated aragonite needles. Again, etching in dilute HCl is generally required to reveal them—more proof that they are not laboratory artifacts, but are contained within the rock. At Viterbo they are abundant in crystals of native sulfur and this fact led to their eventual discovery in metallic sulfide minerals such as pyrite, chalcopyrite and chalcocite. Bacteria had long been suspected of playing a role in the deposition or beneficiation of ore minerals, but no one had looked for the presence of nannobacteria. In most sulfides they are revealed best by etching in HCl or HNO_3.

Silica minerals also show evidence of precipitation by nannobacteria. Such has been observed to be the case with opal, chalcedony, chert and cristobalite. They are revealed by brief etching with HF. Again, some cherts consist of very closely packed nannobacteria, in other samples they are scanty or even absent.

Even complex silicates as clay minerals appear to have been produced by nannobacteria. The bacterially formed clays occur either as filaments with nannobacterial beads of the same size at their tips (like the head on a matchstick), as clumps of tiny balls or as swarms of caterpillar-like bodies. In clays the nannobacteria can range as small as 0.03 micrometers.

Preliminary studies imply that nannobacteria have an active role in the rusting of iron, the "greening" of copper, and the solubilization of metallic aluminum. They also contribute to the plugging of pipes by mineral "scaling," and appear to be involved in the construction of the $CaCO_3$ shells of clams, foraminifera, and even birds' eggs.

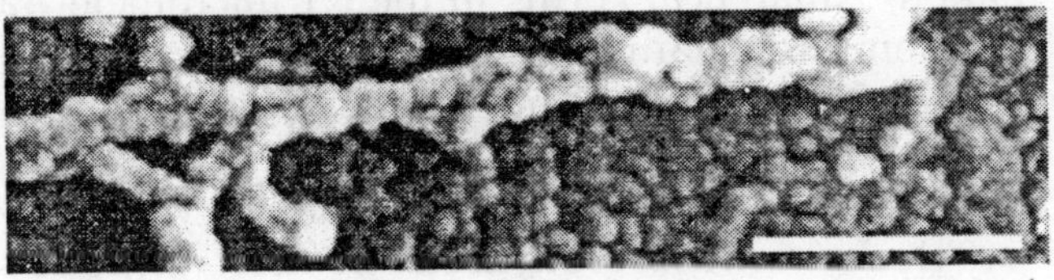

Fig. The nannobacteria.

A piece of rusted iron from Port O'Conner, Texas. This was treated with HCl for 10 minutes to remove the iron hydroxide, revealing the naked nannobacterial cells that precipitated the iron. The cells cover the iron completely as a sheet of balls, there are a few short chains and one long filament composed of balls. The nannobacteria here mostly range from 0.02 to 0.06 micrometers. Scale bar, 0.5 micrometers.

Problems and Discussion

If nannobacteria are so abundant—possibly an order of magnitude more abundant than normal bacteria, if the abundance by volume in rocks can be indicative—how is it that they could have been missed for so long? Probably the main reason is that microbiologists have little or no interest in the occurrence of any type of bacteria in soils or rocks, and it has been standard microbiological dogma for fifty years that bacteria smaller than 0.2 micrometers cannot exist. This is the smallest size visible with an optical microscope, with which censuses for bacterial populations are taken, and filters of 0.2 micrometers are customarily used under the assumption that they will catch all the bacteria, and the fluid that runs through will be bacteria-free. The few geologists interested in the interaction between bacteria and minerals followed this dogma, and thus never looked at samples with the SEM at such high magnifications—and if they saw 0.1 micrometer bumps, they passed them off as laboratory contamination. A colleague once remarked that nannobacteria might be the "Dark Matter" of the Biological Universe—having enormous effects, but up to now not seen or recognized.

And so the study of nannobacteria chugged along at a sluggish pace until the electrifying discovery of the meteorite from Mars in Antarctica. Forms in the 0.1 micrometer range, exactly resembling nannobacteria found on Earth, were discovered in this meteorite. The "incognoscenti" immediately attacked, using the arguments "too small to be bacteria," "artifacts," being unaware of the previous six years of work done on these minute creatures. In fact, unpublished photographs of work done by F. Leo Lynch on Sicilian volcanic clays with nannobacteria can exactly duplicate every published photograph issued by NASA. Even more "far out," work by us on a carbonaceous chondrite meteorite (Allende) reveals distinct grape-like clusters of 0.05 micrometer balls, probably nannobacteria as well, and this time kosher extraterrestrials without any chance of contamination from earthly ice or soils.

A carbonaceous meteorite currently under study, collectedfrom Allende, Mexico by meteoriticist Elbert King. Grape-like clusters of nannobacterial cells (size 0.02–0.10 micrometers), bridge a pore between two olivine crystals. It is believedthat the majority of the carbon found within carbonaceous meteorites will be found to consist of these balls.

If the idea of the ubiquity and overwhelming biomass of nannobacteria on earth (and perhaps in space as well) is correct, this bears enormous implications for earth-surface chemistry, ore deposition, weathering, in microbiology and even medicine as a potential source of genes or hitherto unseen cause of diseases. Problems of environmental pollution, metal deterioration, and industrial processes should be heavily impacted. If the work on the Martian and other meteorites proves to be correct, extraterrestrial nannobacteria could perhaps be the "Dark Matter" of the astronomical universe as well.

7

Viruses

Viruses may be defined as acellular organisms whose genomes consist of nucleic acid, and which obligately replicate inside host cells using host metabolic machinery and ribosomes to form a pool of components which assemble into particles called VIRIONS, which serve to protect the genome and to transfer it to other cells.

CHARACTERISTICS OF A VIRUS

Viruses are different from anything else found on earth and are mainly characterized by their size, shape, and half alive/half dead existence.

The big difference between viruses and all else, is that fact that viruses are so small they can not be viewed without the help of an electron microscope. This is because viruses are, on average, smaller than a regular wavelength of visible light. In effect, the viruses can hide between light waves, thus making them colourless. They can not be seen by the naked eye or a regular microscope. Viruses are so small in fact, that the largest virus is equal in size to the smallest bacteria. The smallest virus measures only 20 nanometers in length. Because of their incredibly small size, viruses are extremely hard to study and understand.

Shape is also a defining characteristic of viruses. The basic shapes viruses tend to take are rods, filaments, crystals, helixes, polyhedrons and spheres, with added extensions. Almost all human viruses are close to being spherical. Every virus carry proteins and nucleic acids in a protective coat. This protective membrane is called the capsid. Extensions on any virus are called antigens. The antigens allow viruses to identify, attack, and enter its target host. Viruses are not classifiably alive or dead. They seem to be in limbo between each state. Viruses exist this way because they are strictly parasites. That is, they can not survive and thrive without a host or group of host cells. The hosts provide viruses with all the chemicals and molecules they need to survive and reproduce. You might think of viruses' as robots that need to take over a factory to make more of themselves. Without that, the viruses are dormant. Viruses can lie dormant within any host or environment until the proper conditions for their activity are provided. This is why we sometimes say that viruses have incubation periods of certain lengths. Some viruses are also classified as 'persistent viruses'. Such viruses can enter and exit host cells without killing them. Even so, each different virus is stimulated by different conditions and they all have different, specific functions they affect in their host. They are mysterious and dangerous creatures.

EVOLUTIONARY HISTORY

Scientists say, that viruses have existed since the beginning of life on Earth. Viruses have plagued animals, plants, fungi and protozoa as long as anyone can possibly trace back in time. Even as the years passed, and life evolved from a primitive 'primordial ooze' into more complex vertebrates, viruses maintained their reign of terror. But does this mean that viruses haven't changed at all over the millennia? Of course not. Just as life evolved, developed and adapted to the new and changing environment, so did viruses.

In their humble beginnings, viruses existed primarily as unprotected genetic strands that carried hereditary information from newly developed life to its offspring. They were messengers. Over time the ever-changing environment of Earth influenced many changes in the method of this transfer of information. The genetic messengers evolved as their hosts did, and they eventually developed protective outer casings to protect themselves from the elements. As life became more complex and cells began to self-reproduce, viruses lost their primary function. Cells took over the messenger role, and so viruses began to infect rather than exchange genes with their hosts. Almost seemingly set on revenge, the newly evolved viruses infected every living thing, and proscribed each cell with their own genetic formulas. They became parasites. They were unstoppable.

Continuing to progress, viruses developed the ability to jump from species to species by changing their genetic material to fit the new hosts' bodies. They were determined to survive throughout the centuries no matter the cost of life. The viruses of today arc highly complcx and clusive. Over their one million plus years on Earth, viruses have developed their own protection, means of survival and efficient ways of infecting their hosts. The medical researchers of today are constantly studying viruses in hopes that we will soon be able to understand them. But fighting viruses is like fighting an enemy who keeps up with every new advancement in weapons technology; the more time they have, the more precocious and powerful they become.

VACCINES

About 200 years ago Edward Jenner might as well have been known as the luckiest man alive. It was in the year 1796 that this country doctor made one of the most astounding discoveries ever. Of course, at the time Jenner didn't know the magnitude of the medical powers he was experimenting with. The experiment Jenner performed would now be considered extremely crude and dangerous. While practicing medicine in the small town of Gloucestershire, England, he decided to experiment with the effects of cowpox and smallpox. Cowpox was a common occupational hazard in the dairy country of England. Coming from sores on the udders of dairy cows, cowpox was a highly contagious disease with caused fever, nauseas and pustular sores on certain areas of the skin. Based only on an old wives' tale he heard as a teenage apprentice – milkmaids who had been infected with cowpox never became infected with smallpox—Jenner decided to infected his first son with cowpox. A few days later, he infected Ed Jr. with smallpox. His son never got the disease.

With this encouraging result, Jenner decided to infect a young boy, James Phipps, with the contagious material of both diseases. This boy's cowpox infection also healed quickly, and he was back in perfect health after only a short amount of time. Afterwards James was injected with smallpox, but was seemingly unaffected, just like Jenner's son. Although no one at the time understood what exactly had prevented the boy from becoming infected with smallpox, it was certain that this obscure doctor had performed a miracle. Dr. Edward Jenner had discovered the first official vaccine, recorded on the 14th of May 1796. Throughout 1796, cowpox invaded the English countryside, providing Jenner with yet another opportunity to test his promising vaccination theories. He not only began investigating cases of milkers who were protected from smallpox by cowpox, but he also studied other inoculations for diseases such as swinepox and a number of bacterial infections. Jenner went on to publish papers on his experimentation. The paper describing the first discovered vaccination was appropriately titled, "An inquiry into the Causes and Effects of the Variolae Vaccineae, a Disease Discovered in some of the Western Counties of England, particularly Gloucestershire, and known by the name of the Cowpox by Edward Jenner, M.D. F.R.S. & C." Within two years, it was translated into many languages and reprinted all around world. Jenner became famous, but met both good and bad criticism. Newspapers and Magazines mocked his work. They would print cartoons showing vaccinated patients sprouting horns and mooing, with titles like "The Cowpock – or the Wonderful Effects of the New inoculation." Jenner had no idea why the vaccination worked, just that it did, but he spawned the first organized field of viral study. A year before he died in 1823, another great man that would change our view of the world was born.

DISCOVERY OF THE VIRUS

Ninety years after Jenner's first vaccine experiments, a French chemist and renowned microbe hunter, Luis Pasteur, performed a similar marvel. Pasteur, at the time, had been studying the effects of another deadly disease of that time: rabies. He had done a great deal of research with animals, and had begun to notice certain things about infected body tissue. It seemed that as the tissue was transferred from species to species, it became less infective and less potent. Pasteur's theory, was that if this weakened tissue was somehow injected into humans already infected by rabies, that it would protect them from the disease's deadly effects. Pasteur, like Jenner, first tested his vaccination on a young boy, this one bitten badly by a rabid dog. This vaccination was also successful, and the boy remained rabies free for the rest of his life. Pasteur was more conscious of what he was doing than did Dr. Jenner, but he was never able to locate the 'bacteria' that he thought caused rabies.

Still, even with these amazing break-throughs in disease prevention, none of the scientist of the time had any clue to what kind of 'monster' they were dealing with. In 1892, Russian Dmitri Ivanovski discovered the very first clue that set these microbes in a class of there own. Even though he was not in the practice of studying human diseases, Ivanovski gave us the first proof that viruses so exist. Ivanovski's main research included the tobacco mosaic disease. Using special filters Ivanovski attempted to separate out the bacteria that was causing the infection. To his dismay, even after several iterations of the filtering process and exposing it to alcohol and fermalin, the tobacco plants continued to become infected and die.

Six years later, a Dutch botanist, Martinus Biejerinck performed a similar experiment. However, he had not read about Ivanovski's work since it was only published in not very well known Russian journals. He performed the same filtering method, but he took it a step further. Though the filtering method might have removed the bacteria, it might not have removed toxins created by bacteria. These toxins could also cause diseases. To see if it was the toxins, he infected a healthy plant and then tried to infect another plant with fluid from the now infected healthy plant. If it was toxins, the next plant would not be infected. It did however, telling Biejerinck that it wasn't toxins. Trying something different, he let the sap from an infected plant sit for three months and tried to infect a healthy plant. It still infected. He tried adding alcohol and formalin which would be enough to kill microorganisms. It did nothing to prevent the fluid from infecting again. So what was causing this disease in these plants if it wasn't toxins or bacteria? Perhaps it was bacterial spores? They could pass through the filters. To test this, Biejerinck heated the fluid to ninety degrees centigrade. All of a sudden, the fluid stopped infecting the plants! However, this didn't prove it was bacterial spores. Quite the contrary. It takes a hundred degrees centigrade to kill the spores, not ninety. This was something else completely.

Through their research, Biejernick and Ivanovski had discovered a new disease-causing agent. Biejernick believed this agent to be a fluid, he called it a "contagious living fluid." Later this liquid was renamed 'virus' for the Latin word poison.

THE HOST

The average virus experiences a rather dull presence on Earth. In fact, the sole purpose of a virus's existence is to duplicate itself and create more viruses.

There are basically three completely different 'types' of viruses. They are animal, plant and bacterial viruses. Each type of viruses is independent of the others. Meaning that a plant virus can not be transmitted to a bacterium and such. So far, there has never been any type of life found on Earth, which is not susceptible to viruses. Some species have more than one hundred different viruses plaguing their kind.

Within each classification, viruses are specific to a certain kind of cell. Viruses tend to be very picky about their hosts. They also have preferred ways of entrance into their hosts. A virus' method of entry is very specialized, and it is one of the main ways a virus is able to locate it's victims. Take, a virus that targets host cells located in the stomach. If a person inhaled such virus particles they would not be harmed. On the other hand, if any virus molecules were ingested into the stomach, the hosts would immediately be infected. Some viruses even require cells to be in certain stages of life. These viruses may prefer actively dividing cells or cells that are younger.

Viruses use a simple marking process to identify the cells they attack. All viruses have special molecules on the outer covering that can search out and identify particles on cell surfaces. Every cell has a unique set of markers that identifies it to other cells. These surface molecules dictate the cells the each virus can recognize and infect. The interaction between the virus surface and the cell surface determines whether infection of the host will be successful or not. Host cells have to be of a very specific type or viruses will not be able to replicate and survive. All cells have surface receptors, which viruses use to identify the cell by their markers and if they match, attach themselves to it. Both sides have to be in good contact, and conditions have to be just right. When a virus securely attaches itself to a host cell in good condition, the infection begins.

VIRUS STRUCTURE

Viruses are not plants, animals, or bacteria, but they are the quintessential parasites of the living kingdoms. Although they may seem like living organisms because of their prodigious reproductive abilities, viruses are not living organisms in the strict sense of the word.

Without a host cell, viruses cannot carry out their life-sustaining functions or reproduce. They cannot synthesize proteins, because they lack ribosomes and must use the ribosomes of their host cells to translate viral messenger RNA into viral proteins. Viruses cannot generate or store energy in the form of adenosine triphosphate (ATP), but have to derive their energy, and all other metabolic functions, from the host cell. They also parasitize the cell for basic building materials, such as amino acids, nucleotides, and lipids (fats). Although viruses have been speculated as being a form of protolife, their inability to survive without living organisms makes it highly unlikely that they preceded cellular life during the Earth's early evolution. Some scientists speculate that viruses started as rogue segments of genetic code that adapted to a parasitic existence.

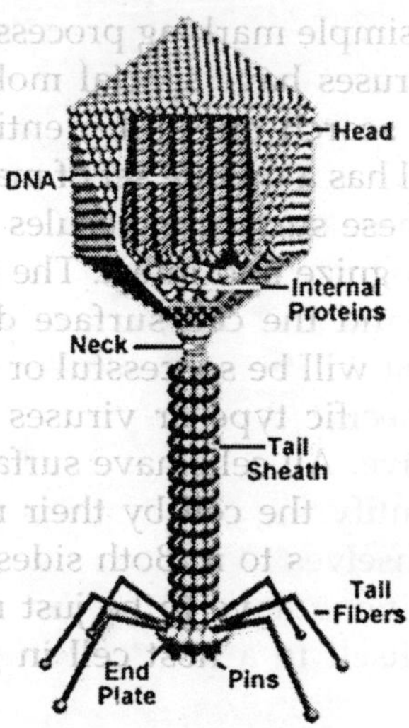

Fig. Bacteriophage Structure.

All viruses contain nucleic acid, either DNA or RNA (but not both), and a protein coat, which encases the nucleic acid. Some viruses are also enclosed by an envelope of fat and protein molecules. In its infective form, outside the cell, a virus particle is called a virion. Each virion contains at least one unique protein synthesized by specific genes in its nucleic acid. Viroids (meaning "viruslike") are disease-causing organisms that contain only nucleic acid and have no structural proteins. Other viruslike particles called prions are composed primarily of a protein tightly integrated with a small nucleic acid molecule.

Viruses are generally classified by the organisms they infect, animals, plants, or bacteria. Since viruses cannot penetrate plant cell walls, virtually all plant viruses are transmitted by insects or other organisms that feed on plants. Certain bacterial viruses, such as the T4 bacteriophage, have evolved an elaborate process of infection. The virus has a "tail" which it attaches to the bacterium surface by means of proteinaceous "pins." The tail contracts and the tail plug penetrates the cell wall and underlying membrane, injecting the viral nucleic acids into the cell. Viruses are further classified into families and genera based on three structural considerations: 1) the type and size of their nucleic acid, 2) the size and shape of the capsid, and 3) whether they have a lipid envelope surrounding the nucleocapsid (the capsid enclosed nucleic acid).

There are predominantly two kinds of shapes found amongst viruses: rods, or filaments, and spheres. The rod shape is due to the linear array of the nucleic acid and the protein subunits making up the capsid. The sphere shape is actually a 20-sided polygon (icosahedron).

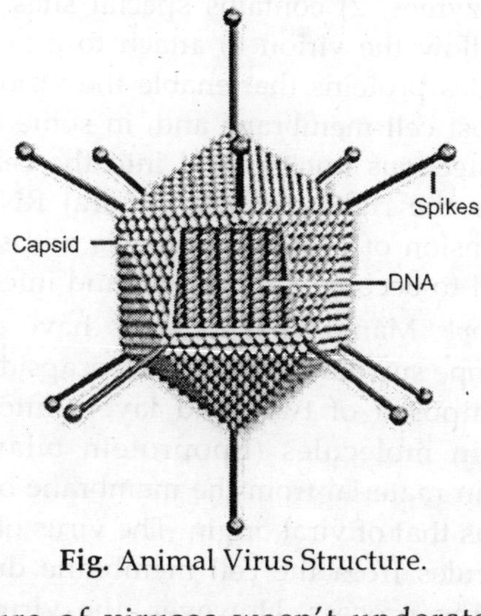

Fig. Animal Virus Structure.

The nature of viruses wasn't understood until the twentieth century, but their effects had been observed for centuries. British physician Edward Jenner even discovered the principle of inoculation in the late eighteenth century, after he observed that people who contracted the mild cowpox disease were generally immune to the deadlier smallpox disease. By the late nineteenth century, scientists knew that some agent was causing a disease of tobacco plants, but would not grow on an artificial medium (like bacteria) and was too small to be seen through a light microscope. Advances in live cell culture and microscopy in the twentieth century eventually allowed scientists to identify viruses. Advances in genetics dramatically improved the identification process.

- *Capsid*: The capsid is the protein shell that encloses the nucleic acid; with its enclosed nucleic acid, it is called the nucleocapsid. This shell is composed of protein organized in subunits known as capsomers. They are closely associated with the nucleic acid and reflect its configuration, either a rod-shaped helix or a polygon-shaped sphere. The capsid has three functions: 1) it protects the nucleic acid from digestion by enzymes, 2) contains special sites on its surface that allow the virion to attach to a host cell, and 3) provides proteins that enable the virion to penetrate the host cell membrane and, in some cases, to inject the infectious nucleic acid into the cell's cytoplasm. Under the right conditions, viral RNA in a liquid suspension of protein molecules will self-assemble a capsid to become a functional and infectious virus.
- *Envelope*: Many types of virus have a glycoprotein envelope surrounding the nucleocapsid. The envelope is composed of two lipid layers interspersed with protein molecules (lipoprotein bilayer) and may contain material from the membrane of a host cell as well as that of viral origin. The virus obtains the lipid molecules from the cell membrane during the viral budding process. However, the virus replaces the proteins in the cell membrane with its own proteins, creating a hybrid structure of cell-derived lipids and virus-derived proteins. Many viruses also develop spikes made of glycoprotein on their envelopes that help them to attach to specific cell surfaces.

- *Nucleic Acid*: Just as in cells, the nucleic acid of each virus encodes the genetic information for the synthesis of all proteins. While the double-stranded DNA is responsible for this in prokaryotic and eukaryotic cells, only a few groups of viruses use DNA. Most viruses maintain all their genetic information with the single-stranded RNA. There are two types of RNA-based viruses. In most, the genomic RNA is termed a plus strand because it acts as messenger RNA for direct synthesis (translation) of viral protein. A few, however, have negative strands of RNA. In these cases, the virion has an enzyme, called RNA-dependent RNA polymerase (transcriptase), which must first catalyze the production of complementary messenger RNA from the virion genomic RNA before viral protein synthesis can occur.

VIRUS LIFE CYCLE

The life cycle of viruses differs greatly between species but there are six *basic* stages in the life cycle of viruses:

1. Attachment is a specific binding between viral capsid proteins and specific receptors on the host cellular surface. This specificity determines the host range of a virus. The human immunodeficiency virus (HIV) infects only human T cells, because its surface protein, gp120, can interact with CD4 and receptors on the T cell's surface. This mechanism has evolved to favour those viruses that only infect cells that they are capable of replicating in. Attachment to the receptor can induce the viral-envelope protein to undergo changes that results in the fusion of viral and cellular membranes.
2. Penetration: following attachment, viruses enter the host cell through receptor mediated endocytosis or membrane fusion.
3. Uncoating is a process that viral capsid is removed is degraded by viral enzymes or host enzymes thus releasing the viral genomic nucleic acid.

4. Replication involves synthesis of viral messenger RNA (mRNA) for viruses except positive sense RNA viruses, viral protein synthesis and assembly of viral proteins and viral genome replication.
5. Following the assembly of the virus particles post-translational modification of the viral proteins often occurs. In viruses such as HIV, this modification, (sometimes called maturation), occurs *after* the virus has been released from the host cell.
6. Viruses are released from the host cell by lysis. Enveloped viruses (e.g., HIV) typically are released from the host cell by "budding". During this process, the virus acquires its phospholipid envelope which contains embedded viral glycoproteins.

HOW VIRUSES INFECT

Viruses do not possess any life sustaining characteristics, and do not require any nutrients. In fact, without a proper host viruses lie dormant indefinitely. Infection takes place when a virus comes in contact with its intended host. As soon as a virus encounters its victim, it attaches itself to the organism. Furthermore, most viruses prefer a certain type of host cell and a specific mode of entrance. Naked viruses, those without a structured casing, directly enter the cells while other types of viruses fuse themselves to the outsides of their victims and inject their genetic material inside the cell. Once the genetic material of a virus is transferred to the host cell, the virus can 'take over' by incorporating its DNA into the hosts DNA much like they used to do in the prehistoric days. The infected cell is essentially a factory in charge of virus manufacturing. In a process called budding, mature viruses leave the cell a few at a time. Lysis is the much more devastating cousin of budding. In lysis, the cell membrane of the host is completely destroyed, killing the cell. The new viruses are unleashed instantaneously. Almost all viral infections result in the death of the host, but in rare cases viruses leave their host cells alive. When this happens the cells are normally damaged beyond repair. With each successive transmission between hosts a virus is able to replicate itself thousands of times, and ensure the continuance of its reign of terror.

TYPES OF INFECTIONS

Most people relate viruses with 'the cold' and 'the flu,' viral infections that bring out a familiar set of symptoms, and then leave as quickly as they show up. These are examples of acute infections. Acute infections cause many of the minor illnesses that humans have experience with. This type of viral infection is, on the most part, rather harmless causing discomfort and minor indications of cell damage. Acute infections, however, can become much worse if they are recurring and do less damage to host cells. In this case acute infections become chronic infections. Chronic infections can be dangerous and sometimes deadly. In cases of chronic infection host cells may not be damaged at all, but their functions may be disturbed. This can cause serious recurring illness and disease. In many cases, chronic infections can be monitored and the viruses that cause them can be cultured in labs. However, scientists can not culture the types of viruses that cause latent infections. Latent infections come from viruses that can manage to evade attacks of the host's immune system. Latent viruses are persistent and frequently cause deadly diseases.

WHAT'S AN INFECTION?

In normal cells, nucleic acids make up the all the genetic material found in the nuclei. Bundles of nucleic acids are collectively called genes, and they stick together to form chromosomes. There are a total of 46 chromosomes in every normal human cell nucleus. The smaller packets of nucleic acids make it possible for normal cells to manufacture proteins, enzymes, energy and heat. They also allow cells to carry out the everyday functions of life and pass on their characteristics to their offspring. If we take the time to inspect these nucleic acids more closely, we would see that they are specialized molecules known as DNA (deoxyribonucleic acid) and RNA (ribonucleic acid). These molecules are the basic building blocks of all life and act as the master plans for all the cells in a body.

While DNA can be considered the director of operations, RNA specializes in carrying 'messages' from the nucleus to different parts of the cell. DNA is made up of many separate fragments that each contains different coded messages. In a process called transcription, a rough copy of the DNA's gene is decoded and written onto a strand of RNA. The result is a form of messenger RNA, which travels to each part of the cell carrying specific instructions and commands. This process is vital for the survival of cells in a body, and if something hampers its completion or alters the instructions, the results can be deadly.

Viruses are tiny packages of genetic material without a living cell enveloping them. The key to their power is the nucleic acid they possess. When a virus attacks and infects a vulnerable living cell, it pours its own DNA and/or RNA inside. Once inside, the hereditary material begins a virtual Coup d'état. It attaches itself to the cell's existing DNA and sets up a new command system. Now, instead of producing substances the cell needs to survive, it is forced to produce viral nucleic acid. One cell can be used to create thousands of new, mature viruses. The fastest virus only needs 24 minutes to explode a cell and release new virus particles. Cells are damaged and destroyed with each new birth, and chaos is all that is left in the wake.

RNA/RETROVIRUSES

The majority of viruses force hosts to do their bidding by regular DNA to RNA transcription. There are, however, other types of viruses called RNA viruses that follow a slightly different method. RNA viruses such as polio, are made up of messenger RNA instead of the regular DNA. They skip the step of incorporating their genetic material into the cell's DNA. Instead, they pretend they are part of the cell's regular messenger RNA and directly pass messages to other parts of the cell to start making viruses.

Usually, viruses follow this method:

1. Incorporate DNA into cell's DNA
2. DNA makes messenger RNA
3. Messenger RNA directs the rest of the cell to make viruses.

However, RNA viruses skips some steps:

Messenger RNA directs the rest of the cell to make viruses.

Another type of RNA viruses called retroviruses are only able to take control of these cells in reverse. Retroviruses contain a shorthand code of directions coded on RNA strands. For these viruses to be able to replicate and produce protein, they have to convert RNA messages to DNA. Retroviruses contain a special enzyme called reverse transciptase, which allows this process to take place. Once the coded messages are translated into DNA, the virus' message is incorporated directly into the cell's genetic code.

Retroviruses follow these steps:

1. RNA changes to DNA
2. DNA is incorporated into cell's DNA
3. DNA makes messenger RNA
4. Messenger RNA directs the rest of the cell to make viruses.

HIV, the virus that causes AIDS, is one of the better known retroviruses. Some scientists theorize that RNA viruses are the cause of cancer. Although it is not yet proven, this solution would explain how the form and function of cancerous cells are affected.

BEYOND VIRUSES: VIROIDS

For the longest time, virologists were fairly certain that there was nothing on Earth comparable to viruses. Dr. Theodore Diener made an amazing discovery, which questioned these old theories. Diener had accidentally stumbled upon a new and exotic infecting agent. This new microorganism had the same devastating effects as viruses, but lacked many of the properties that define viruses. Viroids are found only in plants and are believed to be a more primitive version of the ordinary virus. In fact, viroids may be living fossils of the genetic messengers which existed before animal cells were created. Viroids are naked strings of amino acids with absolutely no covering of their genetic material. They are free-floating, single stranded RNA. Viroids are known for their simplicity. Once thought to be the smallest infectious agents, Viruses are now considered larger and more complex than viroids. There was great skepticism concerning his discovery, but Diener was eventually able to prove the existence of the viroid, the lethal younger brother of the virus.

VIRUS RESEARCH

Everyday, virologists study harmful viruses trying to understand exactly what these mysterious beings are, and how they behave. The goal of their research is simple. They want to find ways to prevent control and ultimately cure many of the deadly viruses that plague life on Earth. Already their work has been able to prevent several epidemics and wipe out the threats of many viruses. In fact, many vaccines can already control viruses such as smallpox, yellow fever, polio and influenza. The most dangerous viruses, including AIDS and certain brain-destroying viruses are of the highest priority right now. An incredible amount of time and money is being poured into the search for medications and vaccines to help solve their devastating effects. A drug called AZT is one important result of these studies. AZT blocks virus replication in cells and has helped many AIDS patients to resist infections. Further studies are also being done to try and connect viruses with cancer.

By studying viruses directly, scientists have been able to learn a lot about ourselves in the process. We have discovered an incredible amount of information about the actual DNA, RNA and genes of living beings. Once they are able to identify the life controlling functions of specific genes, they may be able to design special drugs to target virus vulnerabilities.

Unfortunately, development of drugs cost biotech companies millions of dollars of research and testing. This limits the types of disease that are being studied. Diseases that predominately affect the poor are often ignored because they are not "profitable" to look at. However, diseases that affect the poor are the ones that are of the greatest concern. These are the diseases that affect the majority of the human population. Thus, the irony of the whole situation.

We've come a long way since Jenner and Pasteur. With our new technology and methods of experimentation, scientists aim to ultimately strip viruses of their mystery, and learn how to prevent the disaster they cause.

General Prevention of Viral Infections

There are no miracle drugs against viruses. However, there are a few antiviral that have been developed, but they don't come close to being as effective against viruses are antibiotics are against bacteria. Vaccines are currently the best medical option available in preventing and treating viral infection. We are still far away from finding effective medical treatments against these diseases.

The good news is that most of the effective preventive measures aren't related with medicine at all. The best way to prevent blood related viruses such as HIV is to:

1. Practice safe sex.
2. Don't be involved in sharing needles.
3. Avoid having contact with an infected person's blood without proper protective measures.

Plus, the biggest advancement we have against fighting any disease, not just viral ones, comes from the invention of the toilet. Sanitation is the best method we have of protecting humans. Most of these diseases plague poor third world countries where sanitation is not very good. While most of the countries are relatively sanitary, with our sewer systems, clean water, etc., parts of this country as well as other industrialized nations are also faced with similar sanitation problems. As long as these places exist, viruses and other diseases will continue to be a problem for the human race. These Hidden Killers will continue to stalk.

IMPORTANCE OF VIRUSES

Because viral processes so closely resemble normal cellular processes, abundant information about cell biology and genetics has come from studying viruses. Basic scientists and medical researchers at university and hospital laboratories are working to understand viral mechanisms of action and are searching for new and better ways to treat viral illnesses. Many pharmaceutical and biotechnology companies are actively pursuing effective antiviral therapies. Viruses can also serve as tools. Because they are efficient factories for the production of viral proteins, viruses have been harnessed to produce a wide variety of proteins for industrial and research purposes. A new area of endeavor is the use of viruses for gene therapy. Because viruses are programmed to carry genetic information into cells, they have been used to replace defective cellular genes. Viruses are also being altered by genetic engineering to kill selected cell populations, such as tumor cells. The use of genetically engineered viruses for medical intervention is a relatively new field, and none of these therapies is widely available. However, this is a fast-growing area of research, and many clinical trials are now in progress. The use of genetically engineered viruses extends beyond the medical field. Recombinant insect viruses have agricultural applications and are currently being tested in field trials for their effectiveness as pesticides.

8

Effect of Fungi on Plants

Plants are the principle source of energy for the biosphere. Thus it is not surprising that all heterotrophic organisms are linked either directly or indirectly with plants, using the energy fixed during photosynthesis. Fungi interact with plants in various ways. Some of these interactions cause obvious responses in the host plant. Symbionts such as endophytes and mycorrhizal fungi modify the plants environment to benefit the plant. Pathogens simply exploit the plant. These interactions are complex. Not only does the plant function as a source of energy, but the plant may provide signals that enable symbionts (and pathogens) to recognise a compatible host. The plant also recognises signals supplied by the fungus. Thus the process of colonisation is a series of steps. At each step, the path may be enhanced or blocked. Even in an established colony, the interaction continues. Fungi and plants continue to communicate and respond.

MYCORRHIZAS

The most common type are arbuscular mycorrhizas, which are characterised by the presence of arbuscules in cortical cells of the root. The arbuscules are linked by either intercellular or intracellular hyphae within the root to mycelium in the surrounding soil. Some fungi form rounded storage structures called vesicles in the root tissue or soil.Ectomycorrhizas are characterised by a sheath or mantle of hyphae surrounding the root. Lateral projections of the mantle may penetrate between epidermal cells of the root in what is called the Hartig net.

Ericoid mycorrhizas are characterised by the penetration of epidermal cells of the fine hair roots by fungi. Coils are formed in these cells and the fungi do not spread from cell to cell.Orchid mycorrhizas are characterised by hyphal coils in cortical cells of root-like tissue or the collar of the shoot. The fungus penetrates through passage cells of the exodermis and hyphae ramify through the tissue by intracellular penetration. Coils collapse over time. Cells may be colonised by more than one hypha on more than one occasion.

Ectendomycorrhizas are characterised by the formation of a sheath on the root surface, with intracellular penetration of and coil formation in cortical and epidermal cells beneath. A Hartig net may be present. Other types of mycorrhiza have been found, often in single plant species or genera with fungi known for form typical mycorrhizas in another plant group. This indicates that the function of the mycorrhiza may differ so the associations are covered a separate entities.

INTRODUCTION TO FUNGAL ENDOPHYTES OF PLANTS

The plant is an extraordinarily common source of organic energy. It is thus likely that a huge array of fungi interact with plants. Of these, a range can be isolated from apparently healthy tissue, and many have never been documented to be associated with disease. Others may cause disease when environmental conditions change. Others appear to be the organisms usually associated with initial degradation of the leaf following senescence.In addition, one specific group of fungi have a close association swith grasses. These fungi are transmitted vertically (through seed) and their impact on the economic value of the host has led to a study of their biology. The grass endophytes and the related fungi are considered separately from the more common general endophytes.

A wide diversity of fungi are isolated from the tissues of most terrestrial and marine plants, and red and brown algae. Fungi are present in most plant parts, especially the leaves.

Where the tissue is apparently healthy, the fungi may be either endophytes, epiphytes or latent pathogens. Endophytes are contained within the plant without disease. Plant tissues remain entire and functional.

Endophytic Fungi

Endophytes colonise plant tissue and remain within the tissue except that fruiting structures may emerge through the surface. Indeed, leaves may be fully colonised by a variety of fungi within a few weeks of emergence. The colonies remain asymptomatic and some in perennial parts may have a very long life. Endophytic fungi are found in all divisions of fungi so have presumably evolved the association independently on many occasions. The most common endophytes are anamorphic members of the Ascomycota, and they are often closely related to fungi known to cause disease, either in healthy tissue or as secondary invaders of damaged tissue. This suggests that the endophytes may have evolved from pathogens or vice versa. The mechanisms of host recognition and development of colonisation may also be common. A wide range of plants have now been examined for endophytes, and endophytes have been found in nearly all of them. An enormous number of different fungi can be isolated from plants growing in their native habitat. Most of the fungi are uncommon and narrowly distributed, taxonomically and geographically. However a few fungi are widely distributed with the host, suggesting a long standing, close and mutually beneficial interaction.

Indeed, some fungi are found in many different terrestrial hosts, especially endophytes of crop plants. While most information has been gathered from terrestrial ecosystems, fungi are found in algae and seagrasses. Just as we know less about marine ecosystems, our understanding of the biology of marine endophytes is extremely limited and will not be discussed further.Dispersal of the endophytic fungi remains puzzling. Apart from *Neothyphodium* and related species, endophytes are transmitted horizontally. That is, each plant is colonised by fungal propagules that arrive from the environment. The source of transmission has been determined in only a few cases. Propagules of some endophytes have been found in the body of insect pests of the host. Intriguingly, at least two entomopathogens have been documented as endophytic fungi. Thus insects may disperse the fungi from host to host. Aerial dispersal either in the wind or on vectors is probably the most common mechanism for fungal dispersal. Endophytic fungi colonise various parts of the plant. Many of the fungi sporulate in culture indicating the potential to release spores in the air. Indeed, sporulation is seen after senescence of plant tissues. However, few cases of dispersal have been documented in the wild and the various mechanisms remain unexplored.

Nutrients are cycled between the host and fungus. The fungus clearly gains a predictable environment in which nutrients are readily available. Thus the benefits are clear for fungi establishing endophytic associations. The loss of plant resources to the fungus, and the potential of some fungi to grow rapidly, indicates that the host regulates development of colonies. Each plant host has a range of physical, chemical, constitutive and induced controls over the spread of fungi within tissues. An enormous diversity of phenolic and other deterrent plant compounds are associated with the presence of endophytic fungi, in fact more than are associated with potent pathogens in the same host. In addition, presence of endophytes upregulates plant responses to pathogens. In the absence of plant controls, proliferation of endophytes through tissues would be expected. The reaction of the plant to endophytes suggest that the interaction is one of confinement by the plant. Colonisation by endophytes ranges single cells (*Rhabdocline parkeri*) to patchy distribution through leaves and stems (*Chaetomium globosum*).

The presence of specific endophytes, the lack of signs of disease and the long-lived association indicate that further study of the association may yield interesting information of the precise benefits to plants associated with these colonists. Plants may benefit from the presence of endophytes in many ways. Potential plant benefits have been examined in only a few cases. *Rhabdocline parkeri* produces a compound that reduces needle attack by borers. Metabolites produced by *Phomopsis* sp in cotton appear to deter larvae of *Helicoverpa* from feeding on leaves. The parallels with *Neotyphodium* are clear. In addition, aphids feeding on leaves of cotton may become colonised by *Lecanicillium lecanii*, when conditions permit. Thus the aphid may be killed or it may transfer the fungus to another leaf. In addition, endophytes may upregulate host responses to pathogens and pests. *Chaetomium globosum* has been shown to increase host resistance to rust and tan spot pathogens in wheat. Direct interactions appear to be too small to measure in this case. Presence of *Lecanicillium lecanii* appears to reduce the feeding by aphids from leaves of cotton. The interaction is probaly due to induction of host responses, which is perplexing because plant regulatory pathways for responses to insects and pathogens are not thought to be complementary.

Endophytes appear to have direct and induced effects on plant responses to biotic agents. The interaction with abiotic agents remains largely unexplored.The broader, ecological function of endophytic associations is still being debated. Many fungi that are associated with the initial stages of litter decomposition are found in healthy tissue of the same plants. Thus they are involved in the initial stages of resource recycling. Endophytes are also associated with aquatic activities. Many aquatic fungi have an endophytic stage in their life cycle. An enormous diversity of endophytes are found. Endophytes are probably associated with a wide variety of host functions.

Latent Pathogens

Latent infection of plants by pathogens is the state where a host is infected by a potential pathogen and the signs of infection are absent. Latent infection eventually results in the expression of macroscopic signs. Disease may be induced by changes to the host physiology due to the activity of the fungus, or changes in the environment that increase the stress on the plant. For instance, a decline in available light, or water, or increased herbivory may trigger development of disease symptoms. In some cases, the signs of disease do not appear until the host tissue has reached maturity or senescence.

The causes of disease are possibly due to significant removal of host resources, response to toxins released as secondary metabolites by the fungus, or alteration of host metabolic pathways which reduce host growth rates. At the very least, pathogens would cause dwarfing of the infected host. The period of latency has been defined as the time between initiation of infection to the expression of macroscopic signs. Thus the length of the latency period may be days or years. Latent infections are important for the damage they cause directly. They also weaken the host plant, predisposing it to infection by other pathogens and pests. Further, infection of one plant is likely to spread to adjacent plants as happens with many pathogens of crops. Because the host lacks obvious signs of the presence of latent pathogens, the presence of latent pathogens is likely to go undetected and their effect ascribed to other causes.

Epiphytes

Fungi that exist on the surfaces of plants are called epiphytes. While some fungi are adapted to the plant surface, the community also includes propagules of air-borne species, fungi that would not otherwise be considered epiphytes. The interaction of epiphytes with their host is largely unknown. The surface of the plant, especially the leaf is a challenging environment. Though plants express metabolites through the cuticle, the surface is dry, waxy and affectedby UV radiation. Epiphytic fungi differ markedly from endophytes. Epiphytes have many obvious characteristics that enable continuation through the inclement conditions of the leaf surface. In brief, epiphytes are likely to be coloured (melanised), and thus able to resist UV radiation. Some epiphytes can digest lipidic substrates, and thus may utilise the waxy layer covering the leaf. Finally the yeast form has a relatively short life cycle enabling epiphytes to multiply during the short periods of appropriate environment.

Conclusion

Endophytes include fungi that have one or more of a variety of interactions with their host plant: some fungi are widespread and found on many different plant species; others are highly specific to single hosts in single environments. Further, a diverse array of interactions between plant and fungus have been found. Given that a huge array of fungi may be isolated from any one host, it seems possible that endophytes will have one or more of a wide array of functions, most of which are unknown at present.

Cotton is commonly colonised by Alternaria alternata and A. macrosporum. The fungi are among the first that can be isolated from emergent seedlings. The fungi can be readily isolated from leaf tissue for the rest of the growing season, and spores can be washed from the leaf surface. The leaf tissue is heavily colonised. The fungi rarely cause serious disease. However, when concentration of K in the leaf tissue falls below adequate levels, a purple leaf spot develops which contains Alternaria. Under experimental conditions of K deficiency, inoculation of healthy leaves with Alternaria, results in similar signs. Other factors are also usually associated with development of the disease and causation is apparently multifactorial.

ENDOBIOTICS AND EPIBIOTICS FORMED BY THE CLAVICEPITALES

Included in the Clavicipitales (Ascomycota) are several fungi that form associations with grasses. The associations include a pathogenic epibiotic (outside living tissue) fungus (e.g. Claviceps, Myriogenospora), a partly endobiotic fungus with surface fruiting structures (some Balansia spp), and an entirely endobiotic fungus with no known sexual stages (the agronomically important Neotyphodium). The fungi have the following characteristics:

Claviceps

Claviceps is known from the 15th to 19th centuries where the accidental inclusion of sclerotia with grain used for bread making resulted in neurotoxicological symptoms called St Vitus Dance. The compounds responsible for the symptoms include LSD, which in more recent times has been used as a recreational drug that induces hallucinations. The syndrome, St Vitus Dance, has now largely disappeared in humans because of clean seed. Toxicity is still found in livestock. In a recent case in the central west of NSW, cattle showed signs of toxicosis following grazing of infected pasture grass.

Paspalum found in areas where it is not grazed or mown, often has a sticky inflorescence, which is indicative of infection by Claviceps paspalii. Infected seeds become white scerotia, with the sclerotia bursting from the perianth when mature. The sclerotia fall to the ground when mature and the following spring germinate to commence the parasitic phase of the life cycle. Do not eat this material whatever you do. Claviceps paspalii is toxic. If you handle the material, wash your hands afterwards.

Neotyphodium and Epichloe

In all cases, the fungus contains toxic alkaloids. Thus consumption of the plant poisons the herbivore. Toxicity is most common where grass is short and growing rapidly (eg after rain). The signs of consumption include staggering and unsteadiness (rye grass staggers and fescue toxicosis). Once the toxin has passed through the system, the individual recovers, unless further toxin is ingested. Repeated consumption of the active compounds can have quite severe and long-terms consequences. A group of closely related Ascomycota in the Clavicipitales have evolved a variety of associations with grasses in which the plant is used as a source of organic energy and site of dispersal of the fungi. The fungi also have in common, the formation of secondary metabolites with neurotoxicological effects. Some of the fungi have considerable historical importance, while others have an enormous; though still unrealised, economic importance to the grazing industry.

NEOTYPHODIUM AND THE GRASS ENDOPHYTES

The fungal genus Neotyphodium is known for its association with livestock disorders. In particular, N.

coenophialum in fescue and N. lolii in ryegrass have been intensely studied. This section will examine the mechanisms underlying the disorders of herbivores and the other effects that Neotyphodium species have on plants. Many grasses have associations with Balansioid fungi, the most commonly studied associations are between C_3 grasses (cool season growth) and Neotyphodium. C_4 grasses appear to have epibiotic associations more commonly.

Infected Grasses

Many grasses (Poaceae) in the subfamily Pooideae are associated with intercellular endophytic fungi. The fungi are mainly found in leaves, leaf bases, rhizomes and seeds. The association has no external signs and signs of disease are absent. The association between Neotyphodium and C_3 grasses is highly specific. Fungi isolated from one plant species, or even strain, can be inoculated into a second, where it is unlikely to persist.

Neotyphodium

Neotyphodium colonises the intercellular leaf and leaf base tissue. Usually the fungi can be observed as wavy hyphae located immediately below the epidermis. As the flowering stem elongates, the fungus grows with it and infects seeds as they form and develop. The fungus is transferred asexually from plant generation to plant generation within the seed in what is called 'vertical transmission'. The seed germinates and the fungi colonise the elongating leaf tissues. In other words, the fungus is clonal and the plant host sexual. The fungi have not been found outside the association. The fungi have no known sexual system and no other known mechanism to infect seedlings in the wild.

Feeding Deterrence

Defence chemicals are produced by the fungus, or the host when infected by the fungus. The chemicals disrupt several metabolic processes in herbivores that use the grasses. The active metabolites are found in at least four broad categories:

The ergopeptides ergovaline and ergovalinine are based on lysergic acid amide. Generally known as ergot alkaloids, they induce various metabolic disorders that reduce reproductive rates and growth of grazing animals. The signs include elevated temperature, vasoconstriction, and reduced prolactin in blood. Serum prolactin is a growth hormone that regulates feed intake in mammals. The ergot alkaloids also influence animal gait. The compounds are found in many other members of Balansioid fungi.

Tremogenic neurotoxins Lolitrem A and B are indole diterpenoids. Along with paxilline alkaloids they induce various stiff legged gaits in sheep grazing on infected grass. They also are toxic to or deter feeding by various insects.The unusual Pyrrolopyrazine Peramine is thought to influence insect activity. It is widely found to deter insect feeding. While peramine is the only compound so far chemically characterised, related lipophilic compounds have been indicated in some other endophyte associations. Pyrollizidine-based loline alkaloids reduce feeding by some insects, and acts in concert with the other compounds to reduce feeding and reproductive rates of many more insects. They also appear to be a factor in the allelopathic properties of infected fescue.

The chemicals influence livestock, insects, mites,and pathogenic nematodes. Other compounds are believed to influence fungal pathogens. These chemicals serve as feeding deterrents, reduce rates of reproduction and growth. The effect of the chemical is systemic in the plant; while the fungus colonises only the shoot, the deterrents influence shoot and root feeders. Livestock have reduced rates of growth in cases where endophytic grass is the sole or major source of plant material. The animals tend to develop unsteady gait and stand in water on hot days. The rate of milk production in dairy cattle is also reduced.

Insect attack is reduced in three ways in endophyte infected grasses. Fewer insects land on endophyte infected grasses. Insects will feed for shorter periods and less frequently on infected grasses even when they land and feed. Finally, rates of insect reproduction, specifically number of eggs laid and the viability of eggs, is reduced. Under field conditions, specific populations of fungi and insects have developed, where the toxins produced by the endophyte deplete the insect population, but not completely. The endophyte/grass symbioticum is commonly not adapted to newly arrived insect species. With the arrival of new insects, grazing pressure is quite high until fungal genotypes able to deter the new herbivore become more common in the pasture. Typically, a long established pasture contains a mixture of fungal genotypes.

Drought Tolerance

Endophytic grasses survive under intermittent drought longer than uninfected grasses. The mechanism is unclear, though change in osmotica within the plant tissues following colonisation by the endophyte may be important in some cases. Changes in osmotica suggest that the mechanism is based on changes in hormonal balances, either directly due to the fungi, or indirectly by the fungus influencing plant hormonal balances.

Competition

In natural pastures, Neotyphodium may colonise up to 90 per cent of the grass plants present. Neotyphodium infected plants are more competitive in mixed pastures (eg clover/ryegrass) with uninfected plants. Low levels of infection at sowing (4 per cent) may change to more than 40 per cent within a few years, regardless of whether the pastures are infested with insects, affected by drought or diseased. Established endophyte infected plants have more tillers, and produce more seeds than uninfected plants, regardless of the grazing pressures. The variation between endophyte infected grasses is greater than uninfected grasses and they therefore have a wider region of adaptation than uninfected grasses. However, in soils where mineral nutrients are readily available, uncolonised plants have better survival in competitive interactions. The competitive interaction appears to favour colonised plants, but the mechanisms underlying the resultant population of plants and fungi are complex and interrelated.

Cost of the Endophyte

Published measures of the metabolic cost of the endophyte suggest that it is undetectable. Indeed, endophyte infection may increase phosphate concentrations in the shoot tissue which influences rates of photosynthesis. Nitrogen use appears to be more efficient in some plant/fungus genotypes. The fungus almost certainly produces the deterrent compounds under direction of the host plant. The fungus increases phenotypic variation of the plant. The values of measures of metabolic costs are debatable. Measures in my lab suggest that the cost is variable and determined in part by the density of the fungus, the presence of a mycorrhizal fungus in the root system and the phosphate nutrition of the plant. In many respects the endophyte/grass association is similar to mycorrhizas, with the added input of chemicals that influence interaction with other organisms.

Applications to Agriculture and Turf Culture

Fungi can be selected for production of specific toxins, and selections can be genetically manipulated in culture using classical techniques. The goal of pasture grass breeders is to produce or select fungi that have only marginal effect on livestock but major deterrent on pests and diseases of the grass. Turf farmers have no interest in grazing by mammals, though they too wish to reduce insect damage. So far, reinoculation of grasses with desirable fungi has proven to be a major stumbling block due to the specificity of the association.

The fungus does not survive in seed in hot conditions. In agronomic systems, the fungi may die out if the seed is stored for longer than 6 months, much less if the temperature of storage exceeds 36°C for more than 6 hours. Many so called "endophyte" seed stocks lack the fungus because storage conditions are primitive and seed is often kept for long periods.

Members of a specific genus of fungus release highly toxic metabolites when in association with the grass host. These metabolites have been studied because of their effect on grazing livestock and herbivorous insect pests. The consequence of endophytes, *sensu stricto*, is that they modify the capacity of the host to function in competitive interactions with other plants. This selective advantage enables increases in the proportion of colonised grasses in pastures and presumably native vegetation. It also places herbivores under selective pressure to evolve resistance to the toxins. The wide array of metabolites found in these fungi is typical of all the Ascomycetous fungi intensively studied, suggesting that formation and excretion of metabolically active compounds is a widespread and common phenomenon in fungi functioning to improve the capacity of fungi to survive in association with heterotrophic organisms.

LICHEN

History of Fungi on Colour

During the reign of Julius Ceasar in Rome, the only people allowed to wear purple (Tyrian or Royal purple) were the Emperor and his family. The colour was difficult to obtain, initially because it came from glandular mucus of the shell-fish Murex brandaris and M. trunculus from the Mediterranean. Thus wearing clothes dyed purple indicated great wealth and power. When Roccella, a readily available lichen, was found to provide the colour purple, the use of the royal purple declined.

A purple dye produced by Roccella was reintroduced during the 14 th century in Florence. It was known as orchil, a corruption of the family named Oricellari, who manufactured the dyes. During the 18 th century, another lichen Perusaria was used to produce a purple dye.

Symbolism and ritual were strongly associated with the use of dyes. Colouring the clothes and body constituted an important component of the display. Indeed, while the main trade between Central America and Europe after the Spanish conquest was in minerals, the export of natural dyes was an important second. However, the use of lichens as a source of dyes for cloth has declined. Huge quantities of lichen are needed, often as much as the cloth. It was not until the mid 1800s that synthetic dyes based on analine from coal tar were used commercially. Use of synthetic dyes has dominated the dying industry since then.

Dyes based on fungi, especially lichens, have been used for a long time. The most common colours obtained from fungi are the browns, yellows and greys. These colours are typical of the tweeds of Scotland. Indeed, some Harris Tweeds still require crottle (lichens). The colours of the pH indicator, litmus, and various reds, yellows and mauve are also available especially when using appropriate mordants (fixatives that may also modify the colour). Litmus is a dye also obtained from Roccella. In alkaline conditions it is blue. As the dying conditions became more acidic, the colour changes from blue to purple and then red. These colours can be fixed to the cloth by using an appropriate mordant. Interestingly, alkaline conditions were originally obtained by using ammonia from urine. The urine was kept in a "lake" in which the cloth was dyed. To obtain purples and red, the lake was heated to drive off the ammonia. No doubt the smell was the main reason lakes were maintained away from the houses of the rich. None the less, fungi are a potential source of a wide array of colours.

Not all fungi produce an appropriate colour, and not all fungal colours can be used successfully. To date, rather drab lichens and fruiting bodies of some macromycetes have been used as sources of dyes. Most microscopic forms of fungi have not been tested for their staining properties. Because many fungi can become coloured in culture, they may be a useful source of dye, especially if the dye is required in any great quantity. Some fungi form obvious colours in their sporocarp. However, the colours of many fungi change rapidly or fade during extraction, limiting the value of the commercial product. Other fungi have a colour that is unpredictable. The lack of predictability of colour also limits the use of the dye. The search for new dyes and determination of their properties needs a more scientific approach.

The use of fruiting bodies of fungi and thalli of lichens as a source of dyes is limited by their availability. Unless other forms of production of useful dyes are found, fungi will remain a small and unimportant source of natural dyes for use in industry. Because the dyes can be extracted from thalli, it is possible that cultures grown under appropriate conditions could also become a source of useful dyes. The potential to manage growth conditions during manufacture means that the fungi could be developed for production of specific colours at will. Maintenance of cultures in controlled conditions means that production is commercially more reliable, predictable and manageable.

Now that the cost of petroleum products has increased, the exploration of fungi as natural sources of dyes is increasing. There is also the belief that 'natural' dyes are safer and their production more environmentally sustainable. While use of natural products is still restricted, some interesting products are being (re) discovered. We cannot go back to the past. Lichens are slow growing organisms, harvesting for a large dye industry would deplete reserves rapidly. New approaches to production of natural dyes are necessary.

LICHEN BIOLOGY

Lichens are stable, self-supporting associations between fungi and photobionts. The fungi are most commonly Ascomycota. A few Holobasidiomycetes may be involved in the association. The photobionts are either algae or cyanobacteria. The algae may be single celled or rarely filamentous. In most cases, neither partner is found living alone in the wild.

Lichens are common primary colonisers of stressful habitats. They are major structural determinants of extremely cold environments, and extremely dry climates. Thus, we commonly see lichens in the Arctic and Antarctic, and as crusts on desert soils. They are also seen as the surface layer of trees, rocks, roofing tiles and exposed surfaces in many different environments. In moist environments, the foliar forms are more obvious, being attached to limbs and either growing up or falling down. However, lichens are not found in heavily shaded sites because they require light to support their photosynthesis. In general terms, the photobiont supplies the organic carbon from photosynthesis, and the mycobiont ensures the supply of minerals and water. The fungus most commonly forms the majority of the lichen. In filamentous and gelatinous lichens, the photobiont may form a larger proportion of the lichen. In gelatinous lichens, the cyanobacterium exudes a polysaccharide that functions to absorb and retain water, a function normally attributed to the fungus. The nutritional exchange between partners is probably much more complex than exchange of water and minerals for organic carbon. Cyanobacteria commonly fix dinitrogen. Thus the lichens containing cyanobacteria obtain their organic nitrogen directly. The source of nitrogen in the absence of fixation is presumed to be the air especially following the reduction of dinitrogen by lightening.

While each lichen appears to be a homogenous structure, a single thallus may contain several different species of photobiont and fungus. Even in a single species of fungus, several different genotypes may coexist in the one thallus. Thus, description of the form and function of lichens are generalisations that need not reflect the complexity within a single unit.

Morphology of Lichens

Lichens can be structures that form flat structures that are appressed to a surface, leafy forms that have attachments to their surface or foliar forms that have a single attachment to a surface with the larger part either erect or hanging from the attachment. The fungus primarily determines the form, but the photobiont may also influence it. Thus, complex forms are possible within one lichen.

Crustose: The simplest form of lichen is a crust on the surface. Crustose lichens are highly variable in anatomy. However, they all tend to be adnate or attached directly to their surface. Their growth tends to be radiate, in that the mitotic regions are at the margins, and the centre is more likely to be dying.

Foliose: Foliose lichens have a sheet-like structure, and are attached to their surface by root-like rhizines. The thallus is highly differentiated, with the lower surface being an absorptive tissue and the photobionts being held in a manner that maximises photosynthesis. Commonly, the upper surface is fungal tissue, with the mid-layer containing the photobiont. Growth takes place at the margins, and these tend to be lobed.

Fruticose: Fruticose or umbilicate lichens are attached to their surface by a holdfast. The main body of the lichen is either erect or pendulous, and commonly highly branched. Growth takes place at the ends of the "stems" and may be quite complex.

Evolution: Colonies of rock-surface fungi may be found living in close association with free living algae. In co-culture, the organisms have been observed to initiate a close association. Hyphae branch increasingly as they near the photobiont. The fungus releases mucilage around the contact zone, and close specific contact may result. Interestingly, no evidence of an antibiotic effect has been reported. These observations had lead to speculation that suggests lichens are the result of the interactions between rock-surface fungi and algae. However, many fungi form lichens and few have relatives among the rock-surface fungi.

Anatomy

Most commonly, photobionts are located in a layer within the fungal tissue. The layer is generally oriented in a manner that maximises photosynthesis, and is protected from rapid changes in water availability. Each cell or group of cells of the photobiont is usually wrapped by hyphae, and in some cases penetrated by an haustorium. Moribund cells may be digested by the fungus, but for the most part, the photobiont remains healthy during the functional period of the symbiosis. The increased size of cells of the photobiont indicates that reproduction is regulated by the symbiosis.

The primary photobiont is commonly a green algae. This symbiont is found within a layer below the surface of the lichen. Cyanobacteria may also be held in small eruptions of or under the surface called cephalopodia. Cyanobacteria can fix atmospheric nitrogen, and thus complement the activities of the primary photobiont, which fixes energy. The thallus is commonly layered. The thallus may be covered by or enmeshed in fungal secondary metabolites. For instance, some crustose lichens have a polysaccharide layer on the surface. The photobiont is located at the base of the polysaccharide layer. Polysaccharide layers may also be found within the cortex of the thallus where their function may be different. The thallus is commonly interleaved by hyphal layers. Some thalli have hydrophobic layers on the surface or within the thallus. The hydrophobicity appears to be related to the presence of hydrophobins expressed by the fungus. Finally, the lower layer of crustose lichens lack hydrophobic materials, indicating a role in the uptake of water and solutes for the tissue.

In fruticose lichens, the central core of stems may be hollow. The core may have hyphae oriented in a woven pattern, and the hyphae may be thick-walled and multi-layered. The core may serve a number of functions, including providing strength or stability.The matted anatomy of most lichens is particularly important for uptake and storage of water. Though water can be taken up rapidly, even from condensation at night, it is also lost rapidly. Thus the anatomy is closely linked to the functioning of the thallus. Water is necessary for metabolic processes, and in the absence of water, the lichen slows or stops its metabolic processes.

Reproduction and Dispersal

Lichens may reproduce vegetatively or sexually. The fungi form typical reproductive units. Sexual reproduction tends to result in fungal and algal propagules that following germination must meet with a compatible partner before a functional lichen can form. Sexual reproduction is not considered to be a common means of reproduction to form new units. However, propagules may become entangled with an existing lichen. By initiating growth within an existing matrix, genetic complexity may be added to the unit. Reproduction of basidiomycetous lichen fungi appears to be largely sexual, as these lichens appear to lack specific asexual units.

Asexual reproduction can take place by release of diaspores or vegetative fragmentation. In diaspores, called soredia, small groups of photobionts become surrounded by hyphae, in a crack or zone of the surface called a soralia. The soredia are dispersed when the soralia are opened, either by drying of the lichen surface, or by death of the surrounding tissue. Wind lifts the soredia away from the upper layer, and the propagules are deposited on a new surface.

Isidia are another type of diaspore. These form as elongated outgrowths of the thallus. The isidia break off and can be dispersed as for soredia. Intermediate forms are also found, and isidia may also function as photosynthetic units while still attached to the lichen. The fragmentation of a dried thallus can lead to the dispersal of that organism. This is particularly so with fruticose lichens. Their large thalli are readily shattered and fragments dispersed by wind. Some lichens appear to fragment during drought, resulting in a considerable subsequent increase in lichen units. The fragments have an innate capacity to survive dry conditions. On absorbing moisture, they can regenerate and re-establish a living thallus.

Overall, the patterns of asexual reproduction tend to blend one into the other. Numerous terms have been used to describe these forms leading to some confusion for the beginner. Due to the flexibility of the thallus, the importance of the differences in these forms is limited.

Fungal Partners

Some 95 per cent of lichens are formed by fungi in the Ascomycota. The fungi are paraphyletic, and have been placed in at least 12 orders. The remainder are formed by Holobasidiomycetes. Sexual structures formed by the fungi are similar to closely related non-lichenised fungi, indicating that lichenisation has evolved independently many times.

Secondary Metabolites

Lichens are renowned for the extraordinary diversity of their secondary metabolites. All are formed by the fungal partner. Further, production of secondary metabolites by a basidiomycete partner is unknown, and thus these comments relate to the Ascomycota. Most are formed as part of the Acetate-polymalonate biosynthetic pathway. Compounds formed from the Mevalonic acid pathway and the Shikimic acid pathway are also common. The compounds are detected in the wall or as a layer outside the wall in most cases.

The function of many of these compounds is unknown. However, some have been used in industry, such as for the production of litmus, and others appear to have taxonomic importance. Further, the distribution of compounds may change within the thallus. Compounds commonly extracted from the surface are not necessarily found in the middle layers or at the base of any one thallus. Chemotaxonomy has been used for a long time in lichens. However, the taxonomic significance of the data is still unclear.

Ecology

Lichens play an important role colonising new surfaces. Among the metabolites excreted by some lichens are acids. Acids have the capacity to degrade the surfaces on which they are located, thus releasing minerals for uptake by the thallus. Acidic digestion has the effect of causing the slow disintegration of the surface, especially of limestone and other calcareous materials. "Rusting" of surfaces is probably unimportant in terms of the total uptake of minerals by lichens. Most minerals are extracted from solutes in rain or surface water flow. Lichens have the potential to withstand a wide range of temperatures. Thus the adapt rapidly to local and seasonal changes in temperature and they are found in bleak artic and desert environments.

Lichens also grow extremely slowly. Thus any one thallus may be many decades old. The outer edge is probably the only active component of the thallus, unless the lichen has started to overgrow itself. The inner part is commonly inactive.

The thallus also has the capacity to cope with the frequent aridity of their environment. Foliose thalli will curl as the thallus dries, and then flatten as it rehydrates. Photosynthesis follows the pattern of wetting and drying. While changes in form assist enable a return from dehydration, the presence of trehalose, and possibly a range of polyols, is also important. These metabolites enable the cytoplasm to desiccate, while protecting the functionality of the enzymes. Thus, primary production of lichens is highly dependent on the moisture levels of the environment, but they can survive desiccation. The slow rate of growth and the reliance on minerals in rain or high humidity has consequences for survival of lichens in polluted environments. Lichens absorb all minerals in rain, and the presence of pollutants, including sulphur, will result in the decline of the thallus. Because of their sensitivity to pollutants, most lichens are uncommon in areas affected by acid rain and other aerial pollutants. However, some lichens grow on surfaces containing high concentrations of metals, and must be adapted to those metals. Pollution of a single type is likely to select lichens that can tolerate the pollutant. Changing pollution will remove most. In cities, the pollution profile is variable and changing over time. Thus lichens are disappearing from cities.

Remnants of lichen communities within cities are associated with protected habitats. Church yards for instance may house a wide diversity of lichens. Lichens are not welcome inhabitants of city surfaces, however. The capacity of lichens to "rust" the surface leads to loss of the structural integrity of stone and concrete. Attempts to remove lichens, and prevent the re-colonisation of grave stones and other surfaces, are rarely successsful.

9

Effect of Fungi on Animals

Animals and fungi are heterotrophic. They both require an exogenous source of energy to function. Thus some curious and complex interactions between animals and fungi have evolved. In general, the fungus removes a minute quantity of energy and is thus not a major threat to the animal. However, animals have evolved quite effective immunological responses to deter fungal invasion, indicating fungi have been a severe problem in the past. The presence of fungal pathogens is unusual. Fungi found associated with animals commonly provide a functional benefit to their host. The benefit is commonly a primary or secondary metabolite. The host is able to exist in harsh and difficult environment because the fungus supplies vitamins or digestive functions, or protection through excretion of toxins.

SYMBIOSIS AND COMMENSALS

Symbiotic Interactions

Symbiosis can be defined as the association between dissimilar organisms in which benefit may accrue to both at least for a period during the association. Associations need not be static. In a few cases, the symbiotic phase may continue to a parasitic or commensal phase. The following are examples where symbiosis appears to continue for much of the life cycle of both organisms, and where the functional integration between partners is highly evolved.

Scale Insects and Fungi

Mutually beneficial interactions between insect and fungus can be viewed from the point of view of the partners or humans. A case of the former involves Septobasidial fungi (Basidiomycota) and scale insects. They form an association in which the insect relies on the fungus to provide protection from predators, parasites and pathogens. The fungus gains nutrients and a means of dispersal, and appears to rely on the insect absolutely. The fungus grows over the colony of scale insects, providing a tough barrier. The fungus also penetrates insects through natural openings forming haustoria in the haemocoele of individuals. Insects continue to feed, providing a flow of nutrients to the fungus directly from the plant. Parasitised insects do not reproduce, and other, non-parasitised individuals enable the continuation of the insect, reproducing within the protection of the fungal shield.

The fungus releases a chemical attractant. Once detected, the insect associates with the fungus forming a colony. Some members of the colony become parasitised. The fungus, with access to nutrients spreads over the colony. The insect provides chemical signals that appear to modify the structure of the fungal covering. The fungal shield appears to consist of a series of tunnels, apparently developed in response to the insects. In addition, the fungus is dispersed by scale insects carrying basidiospores from infested trees to uninfested trees.

Leaf Cutter Ants, Termites and Fungi

Fungus-growing ants (Tribe Attini in the family Formicidae) obligately depend on the cultivation of a fungal mutualist for their nutrition. The tribe Attini has been divided into higher and lower groups. The higher Attine ants (the genera Atta and Acromyrmex) cultivate fungal gardens in their colony, using leaf material harvested from adjacent living plants. The fungus found in colonies of higher Attine ants is of the tribe Leucocoprinae in the family Lepiotiaceae.

The symbiosis has been well studied because of the damage ants can do to crops and forests. Found in the tropics of the Americas, Attine ants harvest leaf material from many different plant species, shred the material and feed it to their fungus garden in the ant colony. Each leaf fragment is inoculated with a fragment of fungus and then placed in the fungal garden. The garden consists of chewed leaf surrounded by a dense mycelium. The mycelium is characterised by formation of hyphae with swollen tips called staphylae. The staphylae are harvested by worker ants, held in the gut and subsequently fed to larvae or digested by the adult.

The lower Attine ants have smaller colonies, commonly among leaf litter. They feed their fungi on leaves, insects and other organic matter. The fungi of the lower ants appear to be polyphyletic. Interestingly, sequence analysis suggests that the more stable association between the higher group and Leucocoprinae is also the most ancient.

The fungus forms the majority of the nutrient for larvae, but is relatively unimportant for the adult. However, enzymes in the fungus may play a part is detoxifying or digesting the plant material obtained by the adult. Establishment of new colonies requires transport of mycelium by the queen (vertical transmission) as the fungus is thought to lack sporulation.

Just as in laboratory culture, the fungus garden has a distinct chance of becoming contaminated by unsuitable fungi. Spores of a variety of fungi are common in air, and the ants will possible carry spores on their bodies as they move around their habitat. One fungus, Escovopsis is known to reduce an ant colony very rapidly. Interestingly, the ants of at least one group have a system minimising contamination. Leaf fragments are cleaned before inoculation. Spores licked from bodies and colony surfaces are stored in pouches on the body, and pellets discarded outside the colony. The mycelium is also constantly weeded, removing "foreign" fungi and bacteria. Waste in the nest is taken to a point downhill and away from the mouth of the colony. In addition, the workers that remove waste use an entrance separate from that used for other activities. These approaches reduce the chance of contamination, though heavily contaminated colonies are abandoned.

Interestingly, weeding and grooming are processes that require recognition of 'foreign" and useful symbionts. The recognition molecules or cues are unknown. It might be assumed that as ants use olfactory cues for recognition, that the cue is odorous. Fungi can produce various odours, and each species might have specific cues. However, any "foreign" organism able to mimic a useful symbiont would then be able to benefit from the cleansing and feeding provided to the fungus garden. Production of odoriferous cues remains to be examined in most fungi.

Termites and Fungi

The fungus *Termitomyces* (Tricholomataceae, Basidiomycota) forms similar, specific symbioses with termites in the Macrotermitinae, enabling the establishment of huge colonies in tropical Africa and Asia. In the "higher" termites, the fungus is carried by the queen (vertical transmission) of two species of termite to form new colonies. In the lower termites, horizontal transmission is most likely. Members of *Termitomyces* form basidiocarps and disperse spores horizontally. Thus termites may initiate colonies after selecting fungi from the environment.

The fungus is maintained in a fungal comb within the colony. Premasticated plant material is added to the young comb, and termites feed from the older parts of the comb.

Establishment of symbiosis with fungi by these social insects have enabled the insects to establish large, long-lived and complex colonies. With the symbiosis, the insects are able to occupy niches with abundant resources. The fungi have gained access to the resources and environments where they have limited competition.

Endosymbionts of Arthropods

Microbes in their gastro intestinal tract (GIT) of many organisms aid in digestion. The microbes are referred to as endosymbionts to distinguish them from symbionts that lack a tight association. Apart from the commensal fungi of the Trichomycetes, microbes including fungi are found in the GIT of a wide range of arthropods.

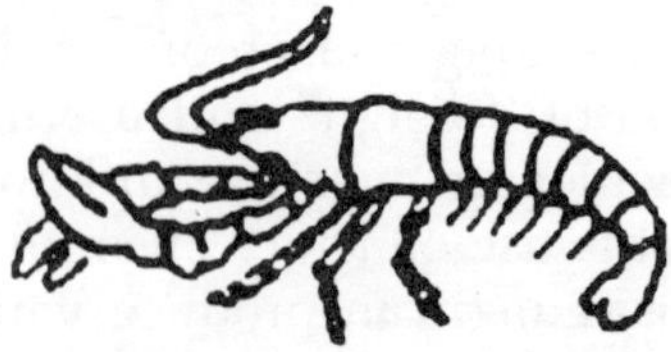

Fig. Diagram of an arthropod.

The microbes are held in various ways. Some are a component of the fluid moving through the GIT. They may have some activity but this is difficult to determine. Others are held in specialised structures within the body indicating the coevolution of the association. The structures include specialised fermentation chambers, and ducts of gastric glands. Endosymbionts may also be held within cells of the epithelium of the gut, Malphigian tubules, or fat bodies. As these different associations are found across a wide range of insect taxa, it might be argued that the association is of considerable benefit to both partners.

The specific activity of fungi is difficult to differentiate from other microbes so the following will consider all microbes. The fungus gains nutrients and protection from the association. The fungus is also transferred from generation to generation of host by deliberate inoculation of eggs from specialised pouches of the insect. Presumably the microbe is ingested as the larvae eat through the shell on hatching.

In most cases, the benefit of the endosymbiont to the host is unknown. In the few cases examined, it appears that the fungus provides its host with nitrogen, which is recycled from ingesta. The microbes probably supply essential amino acids, vitamins and sterols. While some microbes probably have the capacity to digest cellulose, supplementary ingestion of microbial cellulases is likely to be sufficient to degrade the cellulose consumed by the host.

Ambrosia Beetles, Bark Beetles and the Sirex Wood Wasp

A range of insects that are found in living trees utilise fungi they carry with them. The associations tend to be symbiotic, in that the insect carries the fungus from tree to tree either in specific pouches or incidentally. The insect also benefits from the activity of the fungal partner.

Ambrosia Beetles (sub-family Platypodinae) and other related weevils have a highly developed association with a group of ascomycetous yeasts (including Ambrosiella, Ascoidea, Cephaloascus and Dipodascus). The adult insect carries fungi in mycangia, deep pockets in the outer cuticle of the body. The eggs are deposited along with fungi into holes bored into the timber. The fungus germinates and colonises the tunnels bored by the insect. When hatched, larvae feed from the fungus. This association is interesting because the fungi form ambrosia along the walls of the tunnel. Ambrosia are thin walled cells that sprout as a continuous layer from underlying continuous hyphae within the timber. A range of insects feed from the ambrosia, and including the insects that inoculated the timber.

Bark beetles excavate tunnels through the inner bark. Adults and larvae feed on the phloem, and incidentally, any fungi that may be present. The fungi may be carried with the insects from tree to tree. Two pathogens are carried in this way. Both *Ceratocystis* and *Ophiostoma* form long-necked perithecia that exude sticky spores above the bark. These spores are collected accidentally and dispersed to other trees.

When the wood wasp Sirex noctilio arrived in Australia, it caused great concern. The insect is known to cause shoot deformation in Pinus radiata. Females bore into shoots and deposit eggs and the fungus Amylostereum. The fungus causes a white rot of the timber. Spread of the rot is rapid. The fungus digests both cellulose and lignin. Newly hatched larvae feed on the fungus, and also utilise fungal cellulases to digest components of timber. As with Ambrosia weevils, fungal inoculum is carried by the adult female on the outer cuticle, and is deposited into timber during oviposition.

Ironically, a nematode Deladenus may also be carried with the partners. The nematode also feeds off the fungus. On contacting a larval wasp, the nematode then colonises the wasp, in some cases, reducing their fecundity or in extreme cases, sterilising females. When females lay eggs, the fungus and nematode are transferred to the hole. The cycle is thus complete.

A few fungi have evolved close functional and structural associations with insects and other arthropods. The associations are closely integrated and mutually beneficial. In each case, the fungus appears to be gaining access to organic energy, and in a few systems, the animal ensures close association by dispersing with the symbiont attached.

FUNGI OF THE GASTRO-INTESTINAL TRACT

The GIT of herbivores is nutrient rich and oxygen poor. This creates an unusual habitat in which we find a huge diversity of microbes. The microbes have been examined mostly in ruminants of commercial importance. The microbes include a group of anaerobic organisms tentatively placed in the Chytridiomycota that are commonly called "rumen fungi".

Physiologically, rumen fungi are very interesting. They utilise a wide range of enzymes, among them powerful cellulases and xylanases that degrade cellulose and hemicellulose, respectively. As yet the hypothesis that lignin degradation takes place in the GIT remains unsupported. However, these data indicate that rumen fungi have an important role in degrading plant fibre during the digestion of plant material in ruminants and probably most herbivores. Microbes contribute other nutrients to their hosts. Herbage commonly lacks vitamins and essential amino acids. Microbes release these factors either directly or during their degradation. The factors may be absorbed by their host through the wall of the GIT. The role played by fungi, specifically, in this aspect is unclear. The comparative rates of degradation of plant material by fungi alone and in combination with other microbes indicate that digestion is a synergistic process. Release of nutrients by microbes is essential for animal nutrition.

Microbes also detoxify herbage. Plant defence compounds are extremely toxic organic molecules, including alkaloids, diterpenes, and toxins of microbial origin. Microbes remove the toxic components from the food, often using them as an energy source. The resultant food is thus more palatable for the host. The role of microbes in digestion are complex. The presence of oxygen can be extremely harmful to anaerobic organisms. The enzymes superoxide dismutase and catalases are needed to defuse the reactive nature of oxygen, and they are found in all aerobes. Anaerobes lack these enzymes, making culture extremely difficult. Rumen fungi are cultured under strictly anaerobic conditions.

Microbes of the GIT differ in other ways. The anaerobic rumen fungi lack mitochondria. They have hydrogenosomes, organelles commonly assumed to provide energy in the form of ATP. Hydrogenosomes are normally located close to the base of the flagella in the zoospores (shown at right), where considerable energy is used. Little is known about the efficiency of hydrogenosomes. However, energy release from glucose in hydrogenosomes is considerably less than from mitochondria. This suggests that anaerobic fungi, while ideally suited to their habitat, are unlikely to be competitive outside the GIT.

Rumen fungi are found in the GIT of all herbivorous mammals that have been examined, including the marsupials and mammals common in Australia. In the Australian mammals, fermentation may take place in the foregut or hindgut. Animals commonly lack the capacity to digest even small polypeptides and thus rely on their microbial symbionts to complete the task for them. It is likely that similar fungi will be found in herbivores in aquatic environments, and arthropods that utilise plant material. However, some herbivorous insects appear to have very short periods for digestion, and clarification of their mechanisms of digestion is needed. The complex microbial communities of the GIT probably play an essential role in the function of herbivorous animals.

Fungi of the GIT grow by hyphal (rhizoidal) extension. Hyphae enable colonisation of the surface and interior of plant fibres. Enzymic digestion of particles thus is more rapid and complete for the short period plant material is present in the GIT in comparison to bacterial digestion or invagination by protists. The mode of growth also enables colonisation of larger fragments as soon as they enter the digestion chamber. Further physical fragmentation by the animal would result in the funagl population also fragmenting, leading to a more rapid dissolution of the plant material. While the relative importance of fungi in digestion remains conjectural, they appear to play an important part in digestion of cellulosic materials of larger and more complex fragments.

Rumen fungi are assumed to disperse in dung. While the mode of transfer from host to host is unknown, contact between parent and offspring is thought to provide the initial inoculum, and that this is via faecal sources. External transmission means that the fungal propagules must be protected from atmospheric oxygen during dispersal. The mechanism associated with survival of dispersal structures is unknown.

COMMENSAL ASSOCIATIONS

Insects have a wide range of associations with the fungi. However, in some cases, the association, while essential for the fungus appears to have no consequence for the insect host. The taxon Laboulbeniales are widely thought to have a commensal relationship with their insect hosts.

The Commensal Laboulbeniales

A large group of Ascomycota, the Laboulbeniales are found attached to many different insects. The association appears to be highly specialised, in that each fungal taxon appears to be specific to a small group of insects. The fungi colonise the surface of insects, forming a haustorium within the cuticle of the outer surface of the insect. The fungi apparently derive all their nutrition from the host. However, they do not appear to damage their host. The fungi are obligate commensals: attempts to culture the fungi have so far failed.

The Trichomycetes

Members of the Trichomycetes are obligately associated with the internal cuticles, mostly the hindgut, of insects and many other Arthropods. Some argue that the fungus is not a parasite as the fungus does not reduce nutrient supply to the host. Thus these fungi might be commensals too. The distinction between parasites and commensal is difficult to define. Clearly, by accessing organic energy from a host, a fungus may be described as parasitic. As the fungi do not appear to harm their host then they may also be commensal. The entire group remain to be examined more closely, as both the insect hosts and their fungal associations remain poorly understood.

YEASTS AND FROGS

Depending on your point of view, the interaction between fungi and their insect host can be seen as symbiotic if the fungus is found within the gastrointestinal tract, pathogenic if the insect appears to be deleteriously affected, or commensal if the insect appears unaffected. Determining the nature of the interaction becomes extremely complex when other organisms are involved. In this case study, one fungus possibly symbiotic with one host appears to be pathogenic with a second.

Alison Mokany, University of Sydney, examined the interactions between mosquitos and frogs in freshwater. During this research she observed the presence of a yeast fungus Rhodotorula glutinis in her experiments. The fungus lives in the gut of its host frog Limnodynastes peronii where it is probably a beneficial symbiont. The fungus is excreted in faeces into the water by the frog. When in water, the fungus has a direct effect on the rate of development of co-occurring larvae of the mosquito Culex quinquefasciatus. In the presence of excreta from the frog, but not the frog itself, fewer female larvae survive, fewer larvae pupate, and adult females are smaller. The observation indicates that either the yeast, or a metabolic product, acted on female, but not male, larvae of the mosquito. When fungicide was added to the water, female larvae grew normally, which indicates that the yeast may be acting directly. Alison speculated that the yeast causes low-level disease.

The basidiomycetous yeast, Rhodotorula glutinis is found widely in fresh and salt-water bodies, soil and oxygenated sediments, though in high densities only in nutrient-rich environments. Members of the genus are widely recognised as inhabitants of the phylloplane, especially in nectaries and on pollen, where they use simple carbohydrates exuded by their host. Some species are known to be non-fermentative spoilage organisms of stored foodstuffs, including molluscs and fish. Some species of Rhodotorula are reported to grow at very low pH indicating that they can preferentially occupy nutrient rich acid conditions, such as the gut of vertebrates and insects. As they are found in seawater and marine sediments, they can tolerate conditions of low osmotic potential. The gut of marine animals is normally colonised by Rhodotorula in their natural habitat. In their natural environments, Rhodotorula also appears to lack secondary metabolites associated with induction of disease. Rhodotorula can be associated with disease induced by other organisms, though it is unlikely to be the initial causative agent. Thus the aquatic association between larvae of mosquitos and Rhodotorula is very different to other associations of the yeast seen elsewhere.

The reduction in survival, growth and development of female but not male larvae indicates that a specific response to the presence of the fungus is triggered. In different environments, fungi are known to excrete specific signal molecules that are detected by other organisms. Being in water, the molecules would need to be extremely active at low concentrations thus accounting for the response in the experiments reported above, where few experimental animals and insects were used. It seems unlikely though, that males and females would have different responses to the same molecule. It is also possible that the larvae consumed the yeast cells, and that the response of the host was a more direct interaction. The difference between male and female might then be explained by different feeding behaviours: male larvae need not be attracted to the fungus, or may consume different foods. The interaction requires much more detailed examination.

DISEASE

Fungicides

The principle being enunciated in this essay is that fungicides can be used to treat diseases caused by fungi. However, fungi can evolve resistance to the fungicides. In general, resistant fungi either prevent entry, inactivate the molecule, prevent the molecule from docking with the target molecule, or remove the active compound from the cell before it causes damage. Diseases in humans and other mammals caused by fungi have only recently seen to be important. The change in attitude has come about through an increase in tissue transplantation, and the virus HIV, both of involve suppression of the immune system. The first approach to treatment of fungi as pathogens is the use of fungicides.

Many classes of systemic antifungal compounds are currently in use. The action of three will be described and some possible problems with their continued use introduced.

Polyenes

Members of the Streptomyces produce polyene antifungal compounds. The compounds have a broad spectrum of activity, because the compounds are thought to act against the sterol ergosterol. Ergosterol and related sterols are found in the plasma membrane of fungi, where they are important for maintaining structural integrity, membrane fluidity and the function of membrane-bound enzymes.

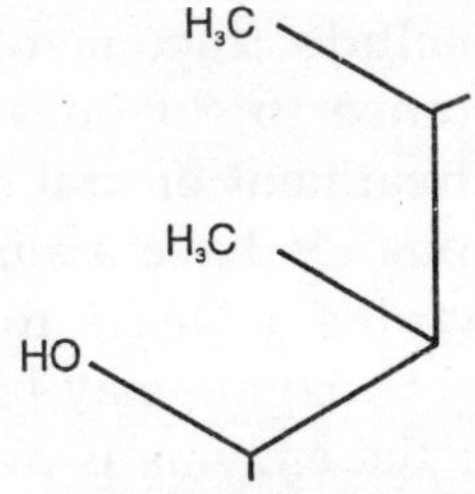

Fig. Structure of Amphotericin B.

Polyenes intercalate into membranes forming a channel through which cell components, especially K ions, may pass. This destroys the proton gradient within the membrane.

Commonly used polyenes include Nystatin and Amphotericin B. They appear to have a greater affinity for ergosterol than cholesterol, thus they target fungal membranes more readily than mammalian. However, their use is not always advised as they may cause greater damage in some tissues such as the kidney, especially with prolonged exposure.

Fungal cells with a lower binding potential to polyenes are resistant to the compounds. The binding potential could be due to fewer ergosterol inclusions in the membranes, or a lower binding receptivity to the polyene.

Azoles

Azoles are synthetic compounds, with two (imidazole) or three (triazole) nitrogens in the 5-member azole ring. They are one of a group of compounds known as Sterol Biosynthesis Inhibitors. Azoles act against ergosterol biosynthesis at the step lanosterol demethylase. The azoles bind to a haem preventing demethylation, and can also be known as demethylation inhibitors. The result of binding is an increase in sterol precursors, with a consequent reduction in membrane integrity and altered structure. Cell multiplication is slowed severely and the altered cells are more susceptible to phagocytosis. Azoles also partially block cholesterol biosynthesis, and thus dosage must balance the action on all sterols in the target tissues.

Azoles in use include fluconazole, itraconazole and ketoconazole. Resistance to fluconazole has been found following long-term treatment of oral candidiasis in people with AIDS. The compounds have a similar mode of action, and resistance to Azoles appears to follow one of two mechanisms. The target enzyme may be modified, such that affinity is reduced, or the enzyme is over-expressed leading to a saturation of the fungicide, with unaltered ergosterol formation. The second mechanism involves increased efflux of the azole. Efflux of azoles may be mediated by one of many enzymes in the plasmamembrane. Efflux of toxic compounds is a common and widespread mechanism resulting in protection of fungal cells. While some efflux pumps targeting azoles have been found, these are unlikely to be the only ones potentially able to reduce concentration of the fungicides within the cell.

As the target site of the fungicide is cytoplasmic, it is also possible that mechanisms either preventing entry or altering the functionality of the plasmamembrane may be found that effectively negate the action of azoles.

Fluoropyrimidines

The fluoropyrimidine 5-FC has a limited spectrum of activity. 5 FC passes through the membrane via a permease. Inside the cell, 5 FC is converted into 5 fluorouracil (5 FU), ultimately being incorporated into RNA. The nucleoside causes disruption of protein synthesis in fungi. While 5 FC has a relatively low mammalian toxicity, 5 FU is a widely used anti-cancer agent. Thus use of 5 FC is limited. Resistance to 5 FC has been reported. Uptake may be reduced if the permease is altered. Conversion to 5 FU may also be reduced if the enzymes involved with the various steps in conversion are modified. Efflux of the compound might also be predicted, as its action is within the cell.

While a range of compounds have been found that act against fungi, problems arise with their use in human disease because of common biosynthetic pathways resulting in low levels of toxicity to the host, resistance due to efflux of the fungal toxin or modification of the target enzyme. The continued use of fungicides, especially in chronic conditions, will increase levels of resistance as has happened to bacterial pathogens.

Human Disease

The principle concept is that diseases caused by fungi are becoming more common. The reasons are complex, but include use of antibiotics that suppress natural competitors of fungi, use of suppressors of the immune system following tissue transplant operations, and infectious organisms, such as HIV, that suppress normal immune responses. In other words, suppression of fungi in the mammalian body is the norm. When immunological mechanisms fail, fungi become a widespread nuisance in humans.

The nature of fungal diseases can be categorised in different ways:

Cutaneous Infection

A group of fungi called dermatophytes can degrade keratin. They exist on the dead skin of humans provided other environmental conditions are met. They become a problem for the host when the skin remains moist and warm for long periods. Each fungus appears to have specific hosts and is found in specific locations in most cases. However, given sufficient inoculum and appropriate conditions, many can swap hosts and locations.

Many of you will be infected by one of these fungi at some time during your life. Tinea and ringworm are the common names given to the disease. The names are usually associated with the place or shape of the infection reaction. The fungus causes distress because the metabolites it releases cause an allergic or immunological reaction. Removal of the infection removes the response.

The fungi are found in three anamorphic genera of Ascomycota; Trichophyton, Epidermophyton and Microsporum. They transfer from host to host usually as infected skin cells. Thus treatment of the condition and removal of the inoculum are necessary for control of epidemics. For instance, flakes of skin may lodge in carpet in bedrooms or on wet floors of change rooms. The fungi rarely cause significant problems.

Candida may also cause surface infection, but is dealt with below where its properties will be discussed.

Subcutaneous Infections

Tread on a spike in bare feet and more than rust and soil may penetrate the tissue. Even an insect bite may result in an infection of fungi normally associated with insects. A variety of fungi may become established in specific situations. The fungus usually changes form on invasion and yeast-like cells proliferate in the body. If the fungus spreads through the body, the spread is associated with the lymphatic system. Spororthrix is a widespread anamorphic fungi that can enter the body through a wound. On entering the host, it converts from a hyphal to a yeast form. The localised infection may resolve itself, but disseminated infection due to spread in the lymph system has been reported. In the worst case, the fungus can spread to joints, bones and internal organs.

Systemic Mycoses

As the name implies, these fungi establish systemic infection in the body. The most common point of entry for the dimorphic fungi is the lungs. The fungi are of two types, those that can establish infection in immune sufficient individuals, and those that require an immuno-compromised host.

Of the former, the usual source of the pathogen is the soil or plant material. Spores are inhaled and germinate in the lung. They then change form and pass into the blood where they are dispersed. Where the specific fungi are normally found, most individuals have antibodies to them. Problems usually only develop when people who have not been exposed come in contact with the fungi. The fungi include Histoplasma capsulatum (Histoplasmosis), Coccidioides immitis, Paracoccidioides basiliensis and Blastomyces dermatidis (Blastomycosis). The fungi are unusual in Australia.

A wide array of saprophytes can initiate disease in immuno-compromised people. Lowered immune competence is associated with involvement of drugs to suppress the immune system such as Cyclosporin A. This drug is now being used widely in transplant operations. The second major cause of immune suppression is disease, especially Acquired Immune Deficiency Syndrome, caused by HIV. The virus, HIV, knocks out the usual response to fungal invasion. Further, prolonged antibiotic treatment, some forms of cancer and prolonged diabetes are associated with reduced immune competence.

Candida

The fungi Candida albicans, various Zygomycota, Aspergillus and Cryptococcus neoformans are especially associated with compromised immune systems. The list of potential opportunistic pathogens is increasing. Candida is a normal occupant of the mucous membranes and GIT of the body. It can cause major irritation when the pH of the mucosa is disturbed, as happens during pregnancy, use of antibiotics and some illnesses. In women it is most commonly noticed as vaginal candidiasis during pregnancy, and in men as balinitis. Young children commonly contract thrush. However, the fungus becomes systemic when the immune system can no longer be sustained. The other fungi are not found among normal human mycoflora. They tend to be ingested or inhaled. The spores germinate, invade adjacent tissues and immediately spread through the blood stream as yeast cells. They have dramatic effects on the host, made all the more dangerous because of the underlying condition. Lack of experience in identifying the fungi among medical practitioners and the lack of sufficient drugs for effective treatment also impede treatment.

NYSTATIN

Nystatin is the name given to the first antifungal antibiotic found to be safe and effective in the treatment of human disease. The compound was first discovered by a New York scientist, Elizabeth Hazen, who, with Rachel Brown, went on to elucidate the structure and activity of the compound. This essay outlines their fascinating story.

Because of the work of Alexander Fleming and Howard Florey, the search for antibacterial agents had reached the point where potentially new compounds were being discovered in microbes extracted from soil. Streptomyces, for instance, had yielded the important antibiotic streptomycin. Actinomyces from soil were generally found to yield relatively high concentrations of a range of different toxins. Bacteria were the usual target for these screening processes. However, once bacterial disease had been suppressed, it became clear that fungi were also important causes of disease. In 1945, no antifungal agents were available to treat fungal disease.

Elizabeth Hazen, of the New York State Department of Health, initiated a screening programme for antifungal agents from microbes isolated from soil. She used the fungi Cryptococcus neoformans and Candida albicans as targets in her bioassays. These fungi were at the time important pathogens becoming increasingly apparent in patients who had been treated with antibiotics that suppressed bacterial competitors. Microbial isolates were grown in either static or shaken liquid culture by Elizabeth Hazen. The bulk extract was placed against the bioassay fungi. Where activity was observed, the process of identifying and characterising the active component commenced. The organic chemist, Rachel Brown extracted from solution and the bulk organism, and fractionated the components. These extracts would then be returned and tested in the bioassay. Some cultures resulted in more than one active ingredient, and sometimes the active component was found in broth and solid materials. Some fractions proved to be toxic, but on purification were found with quite different properties (solubility in water, activity against the bioassays etc). Once active compounds had been found, then they were tested for toxicity against laboratory animals.

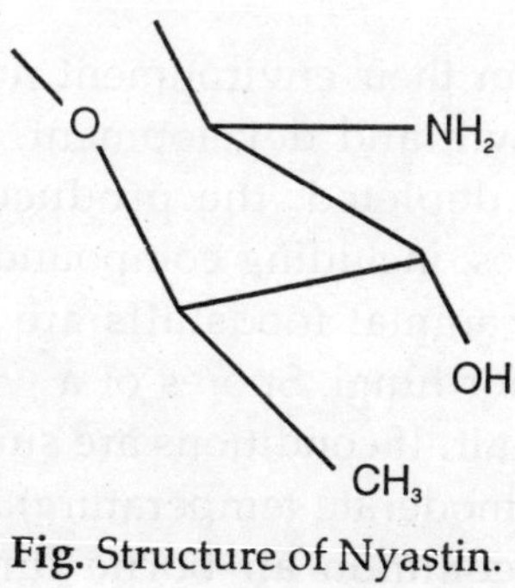

Fig. Structure of Nyastin.

Using this slow and cyclical process, an interesting compound was found in 1948. The compound came from

Streptomyces noursei, a newly discovered actinomycete. Ultimately named nystatin, it was to then pass through a series of processes to first provide enough material to test in humans, and then bulk up for industrial production. Nystatin is relatively insoluble. It can, therefore, be applied superficially, or taken orally for intestinal complaints. It has also been used successfully elsewhere: to protect paintings after inundation, and in plant pathology. Nystatin is similar in structure to amphotericin B, both being polyenes. However, amphotericin B is much more soluble and can therefore be injected for treatment of deep tissue mycoses.

TOXINS

Fungi produce a range of secondary metabolites which include compounds that aid their competitive position. Among the compounds are a variety which cause toxicosis in mammals, especially humans, insects and other herbivores. The compounds of most importance to humans are those which are ingested accidentally with food. Toxins may also be consumed deliberately, either to find out if the mushroom is flavoursome, or because the mushroom has been misidentified. Considering the vast number of toxic fungi, it is surprising that more cases of fungal toxicosis are not observed.

Toxins in Food

Fungi take from their environment nutrients which are used for their growth and development. When the energy resource becomes depleted, the production of secondary metabolites increases, including compounds which are toxic. Human and other animal foodstuffs are a major potential source of nutrient for fungi. Spores of a wide range of fungi are common in the air. If conditions are suitable (access, high water activity and moderate temperature), fungi colonise the foodstuffs. Some common air-borne fungi are known to produce extremely toxic compounds; these include Aspergillus, Alternaria, Fusarium and Penicillium.

Toxins

Common toxins include alkaloids, cyclopeptides, and coumarins. The compounds are active at extremely low concentrations and have a rapid effect. The toxins may cause death. In sublethal quantities, the toxins may also trigger cancer, and influence the physiology of the consumer. Many of the compounds are heat stable remaining active after cooking or treatment of foodstuff. The potential for damage is particularly important for human foods, and food for livestock held in intensive conditions. The most important group of toxins studied for their toxicological effects are the aflatoxins. Aflatoxins are produced by the group of fungi in the Aspergillus parasiticus group, that is Aspergillus flavus and Aspergillus parasiticus. The fungi grow on and in stored grains and animal feeds when humidity is high enough. They also grow in living plants of cotton, peanuts and corn, where colonisation of the host plant may take place prior to seed ripening. The fungi, and probably toxins, are present in the seed at harvest.

The fungi produce aflatoxins, which are difurano-coumarins, in several forms. Aflatoxin B1 and G1 are found most commonly. Their effects are similar. Birds are particularly sensitive to aflatoxins. Feed containing contaminated peanuts given to turkeys in Britain caused widespread death in the early 1960s. The birds died from liver and kidney failure. Aflatoxins also induce cancer of the liver. As the compounds are active at parts per billion, aflatoxins are among the most toxic and dangerous fungal metabolites. Other toxins produced by fungi have attained notoriety. Vomitoxin or trichothecene is produced by Fusarium graminearum and other fungi. The fungus can grow on damp grains, and produce the toxin during storage. It is a potentially important problem for housed animals, such as pigs. This fungus also produces zearalenone, which appears to mimic oestrogen in mammals, and thus causes abortions in pregnant livestock, especially pigs.

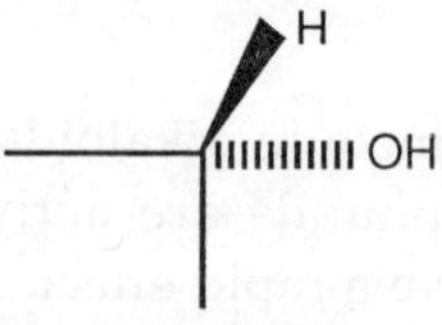

Fig. Structure of different toxins.

Each of these fungi, when closely studied, has been found to produce a range of other toxins, active in different doses, and produced under different conditions. The production of secondary metabolite also varies with strain. Further, some toxins are produced by a range of seemingly unrelated fungi. Unfortunately, toxicology is an area of research which is driven by the end effect of the product, and we do not understand much of the basic biology of metabolite production

Other Metabolites

During the Spanish conquest of the Americas, stories began to filter back to Europe of the use of magic mushrooms, and their place in the cultures of the American Indians. The mushrooms were eventually identified, and subsequently became popular in the Americas during the protests against the Western involvement in the Vietnam War. At this time they were called magic mushrooms, largely because of their hallucinatory effects.

The active ingredients are tryptamine derivatives. They are found widely in two genera, and in a few species of other genera. The fungi are generally saprotrophic, with one species preferentially colonising horse and cow dung. The fungi can be found around the world, and many societies now recognise the use of magic mushrooms, as either a problem or an irritant depending on the cultural conditions. Psilocybin is colourless, and rapidly oxidises to blue, giving the characteristic feature used in field identification. The concentration of active ingredient in mushrooms varies widely, and some unpleasant effects can be triggered unexpectedly. Onset of signs is usually within a few minutes of consumption, and the signs can last for days, though is usually resolved within hours.

Basidiocarps of other species are also known to produce mood alterations, and some have been used deliberately over the centuries, often associated with religious ceremonies. The use of Amanita muscaria is a particularly odd story. Documents suggest that rich of Russia used to consume the fungus for its halucinatory properties. The wealthy used then to urinate into containers for the peasants to drink. Apparently the active molecules used to pass, almost immediately, through the system and were excreted in the urine. Thus both benefited from the drug. However, no other society has used A. muscaria in this way, and in fact the mushroom is more likely to poison the consumer.

Specific Effects of Toxins

A number of fungi produce secondary metabolites which are active in mammals in other ways. Several agaric fungi produce toxins. In Europe, the largest number of poisonings is due to consumption of Amanita phalloides. Amatoxin is found in most species of Amanita, some species of Lepiota, Conocybe and Galerina. Amatoxin is a cluster of compounds, including the peptides amanitin and phallotoxin.

Amanitin is an inhibitor of RNA polymerase II, and thus acts in all eukaryotic cells where protein synthesis takes place. The liver is the first organ in the human body exposed to high concentrations of amanitin after the gut, and is thus the first to be affected.

Phallotoxin was once thought to be responsible for the usual symptoms of amatoxins. The compound acts to inhibit F actin in the cell cytoskeleton. It acts immediately, and probably does not move beyond the lining of the gut.

The activity of amatoxins gives some idea of symptoms of poisoning by Amanita. Amatoxins are taken up by cells lining the gut, where protein synthesis is immediately inhibited. The toxins are released to the blood stream and transported to the liver. In the liver, protein synthesis is immediately stopped. Amanitin is excreted in the urine, and most is evacuated from the body within hours of ingestion.

However, if sufficient liver tissue is affected, liver failure will ensure death.

The dose that is likely to kill an average human is in the range of 6-7 mg, easily found in the cap of one mature A. phalloides. However, like other fungal toxins, the concentration which is fatal for individuals differs, the concentration in different specimens differs, and the environment influences how much toxin can be produced in one basidiocarp.

ANIMAL FEEDING ON FUNGI

Mycophagy

Australia is characterised by a dry climate and nutrient poor soil. Vegetation found in this environment appears well adapted to cope with the stressful conditions. In recent times, examination of macrocarpic fungi has led to the view that the fungi also have adaptations that enable exploitation of the harsh conditions.

Sequestrate Form

Most macrocarpic fungi belong to the Basidiomycota. Their common form in the northern hemisphere is a gilled or poroid "mushroom". This structure enables the dispersal of spores into wind. While this form is also found widely in Australia, many related taxa have rounded fruiting bodies that are found within or below the litter layer (hypogeous). The fungi have retained active discharge of spores, but lack the elevation to release the spores to the wind.

The rounded forms are referred to as sequestrate, and the fungi have probably evolved from closely related epigeous species. The selection pressure for evolution of hypogeous forms appears to be the short periods of moist conditions experienced during fruiting. Hypogeous fungi are found around the world, and occur in the Basidiomycota, Ascomycota, Glomeromycota and Zygomycota. They are particularly common in Australia.

Hypogeous fruiting removes the opportunity to disperse spores in air. Another mechanism is needed if the fungi are to be carried beyond their immediate vicinity. Dispersal is achieved by means of animal vectors. The Black truffle, Tuber melanosporum, has a well characterised chemical attractant that resembles a pig pheromone. Thus the animal (pig) is attracted to the fruiting body, eats it and then defecates the spores at a distant location.

The same pattern of attraction, consumption and dispersal is believed to occur in the Australian hypogeous fungi. Fresh animal digs are seen where the fungi are fruiting. The scats of a range of mycophagous animals contain significant proportions of spores of a variety of fungi. The potential nutritive value of fungi indicates a potential nutritional role of fungi to the mycophagous animal.

Mycophagous Animals

A broad range of smaller animals consume significant quantities of fungus at some time during the year. The animals are mostly small mammals, ranging in size from marsupial mice to small wallabies. They can be grouped into three feeding behaviours. One group, including the Long-footed Potoroo, use significant quantities of fungus throughout the year. Fungi constitute some 90 per cent of their diet. Another group consume significant quantities of fungi for much of the year. The Long-nosed Potoroo and the Tasmanian Bettong are believed to require fungi for most, though not all of the year. A third group consume large quantities of fungi for short periods, often seasonal. Their diet at other times lacks much fungal material. The third group includes rats and bandicoots.

A large range of other animals appears to consume fungi, almost incidentally. These include some birds, wallabies and possums. It is unclear whether the fungi are consumed deliberately, or consumed while eating other materials. In some instances, consumption of above ground fruiting bodies has led to the assumption that the animals are attracted to the potential food source: this mechanism may apply to hypogeous sporocarps.

Taxonomy of Hypogeous Fungi

Interestingly, most of the hypogeous fungi appear to be mycorrhizal fungi, especially ectomycorrhizal fungi. The fungi are commonly found in close association with roots of ectomycorrhizal tree species, including Acacia, Eucalyptus and Casuarina. Hypogeous sporocarps are not found in cultivated soils or areas where the vegetation is highly disturbed. In only a few cases, mycorrhizal status of the fungi has been confirmed by inoculation studies.

The fungi are found in a wide range of taxa. In the Basidiomycota, epigeous fungi are documented in members of the Agaricales, Boletales, Lycoperdales, Ramariales, Russulales, Sclerodermatales, and Stephanosporales totalling at least 230 species. In the Ascomycota, members of the Elaphomycetales and Pezizales total at least 25 species. In the Zygomycota, members of the Endogonales total around 8 species and in the Glomeromycota, Glomales total at least 15 species. With this wide diversity of taxa, it is difficult to make many generalisations about nutritional benefit or functional attributes.

Most of the fungi fruit during autumn in the south eastern and south western parts of Australia. However, some species fruit during autumn and the sporocarps are present for the entire year, and a few form fruiting bodies throughout the year or at times other than autumn. Few mycophagous animals are found in any one area. Thus, providing animals can detect the presence of sporocarps, some food is likely to be available in the woodlands when ever it is needed.

Fire

Only during and following fire is food likely to be scarce. Fire removes the protective litter layer and scalds the surface of the soil, albeit unevenly. A few fungi respond to fire by fruiting abundantly, either immediately or during the following autumn. However, many species of ectomycorrhizal fungi are removed with the litter, and the diversity of fruiting bodies is probably also reduced.

The effect of fire on mycophagous animals is unclear. The size of the population is likely to be reduced directly by fire.

The overall nutritional effect is likely to be related to which fungi are present at the site. Adequate food would be available if species that are protected from and/or respond to fire are abundant. The food supply would be seriously reduced if most fungi are killed by fire. Abundant food supplies would enable a rapid return of the population of mycophagous animals.

Dispersal

The current view is that the fungi benefit from mycophagy through improved dispersal. The basis for the argument is that spores pass through the GIT and are then deposited on the ground a long way from the source. Dung beetles would bury the dung, taking fungal spores into the soil profile. Should roots be actively growing adjacent to the germination of spores of ectomycorrhizal fungi would be stimulated and some fungi would initiate ectomycorrhiza.

If seed were also present in the dung, then on germination roots would be in immediate contact with mycorrhizal inoculum. These speculations require that roots are actively growing in the presence of the buried dung, that germination is possible and that the symbionts are in close proximity. However, there is some evidence that indicates that spores in dung have a very low viability. Further, most animals do not travel beyond the woodland. Roots of woodland plants are likely to be heavily colonised by mycorrhizal fungi. Hyphae from germinating spores will be at a competitive disadvantage with the existing mycorrhizal fungi. When considered with all the information, it could be argued that successful initiation of a mycorrhiza from inoculum in dung is an infrequent event, and that spore dispersal is likely to be a significant factor benefiting the mycorrhizal fungi only with establishing seedlings immediately following severe fire.

NUTRITIVE VALUE OF FUNGI

It is quite clear that hypogeous fungi play an important part in the nutrition of at least some Australian mammals. A great diversity of fungi is consumed and the fungi are eaten for most of the year by at least some animals. Thus they must fulfil the dietary requirements of these animals. However, fungi consist largely of indigestible complex polymers. Unless animals have some mechanism to break apart the fungal structures and break down these polymers, their dietary needs will not be met. It is the importance of these mechanisms of breakdown that enable small animals to utilise the fruiting bodies of hypogeous fungi.

Nutritive Components of Fungi

Few studies have examined the nutritive qualities of fungi. Cork and Kenagy found that fruit bodies accumulated higher concentrations of phosphorus, nitrogen and potassium that seed and leaves. Along with other minerals, fungi also consist of a wide variety of organic compounds, including vitamins, amino acids, steroids and various polymers. Lipids are common storage compounds in the cytoplasm, and the walls consist of complex polymers including proteins and chitin. Thus fungal sporocarps have the potential to be adequate foods for animals with an appropriate digestive system. The potential may be met if the animal can also select different foods that provide its various other requirements.

In a feeding trial, squirrels fed sporocarps of Elaphomyces lost weight. The diet contained adequate total nitrogen and energy. However, the nitrogen and energy were in a form that was inaccessible to the squirrel. Further, a considerable proportion of the spores appear to pass through the GIT unaffected. The study cannot be applied that easily to all animals in the field. One fungus was used and animals would not be expected to subsist on one food source. Squirrels have an omnivorous diet and do not feed solely on fungi normally. The study does indicate some of the difficulties that animals face when their diet is largely fungal.

Little is known of the dietary qualities of fungi. While lipids are likely to be abundant in spores, lipid only becomes available if the spores are broken apart. Protein may be adequate but the composition of amino acids is probably unbalanced and some may be unavailable. The fungi Mesophellia and Rhizopogon appear to have high total nitrogen, though much of it in a protected complex form. Cell walls are also fibrous indicating that they are similar to Elaphomyces. As animals will consume a wide variety of fungi during the year, is can be assumed that they will be exposed to a range of quality and quantity of each dietary need. However, these issues indicate that specific digestive processes are needed to maximise the benefits of consuming fungi.

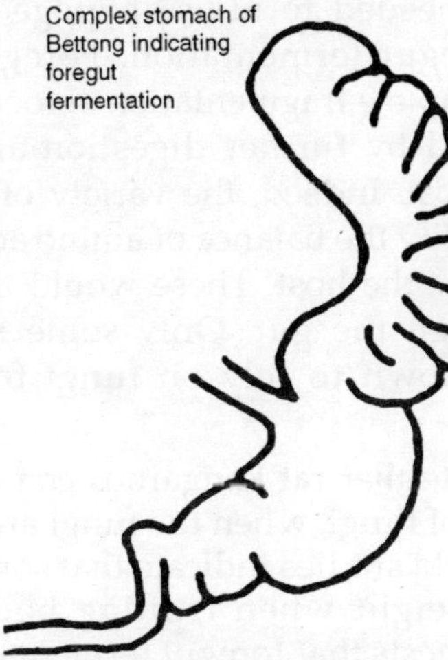

Fig. Mycophagous Animals

Complex stomach of a Bettong indicating foregut fermentation.

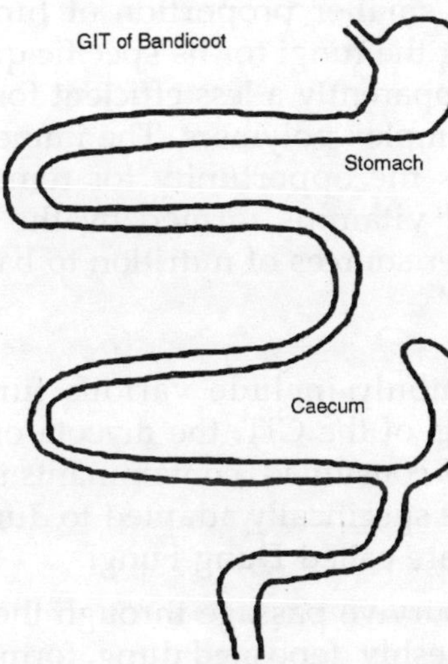

Fig. Diagram of the gastro-intestinal system of a Bandicoot.

Mycophagy is reported most commonly among members of the rat kangaroos (Potoroidae) and rodents (Muridae). Most animals are quite small, mostly weighing between 100 and 3000 g. Of these, only the rat kangaroos appear to have the digestive system needed to utilise hypogeous fungi. These animals have foregut fermentation. Foregut fermentation enables a more complete fragmentation of food, with microbial digestion, followed by further digestion and extraction of nutrients by the host. Indeed, the variety of microbes might be expected to modify the balance of amino acids and produce vitamins needed by the host. These would then be absorbed on passage through the gut. Only some members of the Potoroidae are known to rely on fungi for most of their nutrition.

Studies indicate that rat kangaroos can maintain weight on a limited array of fungi, when the fungi are the only source of food. Further, field studies indicate that some rat kangaroos appear to gain weight when fruiting bodies are readily available. This suggests that foregut fermentation is necessary to maximise the nutritive benefits from the highly protected fungal structures.

In contrast, Rats and Bandicoots lack foregut fermentation. They consume a smaller proportion of fungi in their diet, presumably using the fungi for its specific qualities. Hindgut fermentation is apparently a less efficient form for extracting nutrients from complex polymers. The immediate passage to excretion reduces the opportunity for removal of essential amino acids and vitamins formed by the microbes. Thus rodents need other sources of nutrition to balance their diet.

DUNG FUNGI

Faeces commonly include various fungi. Apart from microbial colonists of the GIT, the mycota of faeces includes fungi accidentally consumed, contaminants settling from air, and fungi that are specifically adapted to dung as a resource. The latter group are called Dung Fungi.

Dung fungi survive passage through the GIT, germinate and grow in the freshly deposited dung, form adhesive spores that are actively and violently discharged. The spores are deposited on vegetation, and are then consumed by herbivores to complete the cycle.

Dung is a very rich medium for fungal growth. It consists of the remains of plant material plus the microbiota associated with its digestion. Much of the material is readily available carbohydrate in addition to the cellulose and lignin. The material is complex, and includes fatty acids, vitamins and amino acids. In addition, initially, the waste is moist, and the pH is close to neutral. It is an ideal medium for microbial growth.

The fungi found growing in dung include those that have passed through the animal, and saprophytes that land after deposition of the faeces. However, most studies of the ecology of dung have noted that the fungi that fruit on the surface consist of a group of recognisable taxa commonly including.

Zygomycota: Pilobolus, Pilaira, Mucor, Phycomyces and Piptocephalis.

Ascomycota: Ascobolus, Saccobolus, Sordaria, Podospora, Chaetomium.

Basidiomycota: Coprinus and Nidulariales.

Energy Required

The order of appearance of sporophores is related to several factors. Apart from environmental conditions, aspects such as energy requirements for sporulation, complexity of fruiting body, and competition have been postulated. Fruiting bodies of Pilobolus and other Zygomycota are simple, and require relatively small input of organic energy. Thus the fruiting bodies should be formed rapidly when energy is abundant and conditions are appropriate.

This hypothesis is based on the notion that Zygomycota can only utilise readily available nutrients, as they lack the necessary array of enzymes to digest complex carbohydrate. Thus these sugar fungi will take up nutrients rapidly and then die back as the nutrients are depleted. They will be replaced by fungi that utilise more complex carbohydrate.

The second postulate is that interspecific interactions determine the order of appearance of sporocarps. The dung is a rich source of nutrients. Fungi that grow will be in strong competition from soon after germination. Fungi will be competing with each other, and the various invaders colonising the resource. Competition will be intense, and we might expect all the aspects of a competitive interaction to be seen in dung.

Specificity

The ecology of dung fungi is still unclear. While some specificity between animals and fungi, and environments and fungi are indicated in surveys, the reason for any specificity is unclear. Further, arboreal animals such as possums and koalas also host dung fungi. As their faeces fall to the ground, and their food is largely above ground, the source of inoculum is unclear.

Spore Release

Release of spores has received considerable attention. The fungi generally expel their spores or sporocarps towards light. In Pilobolus, the release mechanism can target extremely small gaps between surrounding plants. The mechanism controlling this accuracy is a light sensor surrounding the base of the sporangium. When light falls evenly on the sensor, the direction of the sporangium remains unchanged. As the light falls progressively on one side, the sporangium is progressively redirected such that the sensor is shaded.

Similar mechanisms regulate the direction of release of spores of Ascomycota. In perithecia and apothecia, the head of the ascus is pushed above the surface of the ascocarp or through the ostiole of the perithecium. The ascus is directed towards the light where spores are directed during release. By directing spore release, the fungi expel spores towards the gap between plants. Thus the spores are then dispersed away from the fruiting body and towards fresh plant material.

10

Fungi and its Uses

Primary production concerns the commercial production of fungi and fungal products that are consumed by humans or other animals. Industrial production concerns the processes of fungi that may be exploited in any industrial (antibiotic production) or production (biocontrol of pests and diseases) activity.

WEED CONTROL

Plant pathology is an extremely large field of study. The goal of the subject is to understand the field well enough to reduce the damage caused by (mostly fungal) pathogens on economically important plants and plant products. The subject of plant pathology can be turned around, by applying the principles of plant pathology to increase the damage being done to economically important weeds. The basis of the biological control of weeds relies on pathogens attacking the weedy host. The aim of using disease is to poison, or reduce rate of growth, fecundity and general vigour of the host to thereby reduce the weeds competitive position in the crop. In perennials, longevity is also likely to be reduced. Over time, the population of the weed is reduced to a size where it no longer has an economic impact. The pathogen does not eliminate the weed, as that would also eliminate the pathogen.

In fact, the use of fungi to reduce weed populations is extremely difficult. Weedy plants are usually aggressive competitors, growing away from their natural control agents, in an environment that suits their requirements. It is within this severely different environment that we attempt biocontrol of weeds using fungi.

Natural Pathogens or Classical Biocontrol

Early attempts to control weeds took pathogens of the plant from the native habitat of both and introduced the pathogen to the isolated weed population in the new environment. The aim was to establish the pathogen on the weed in the new environment. The approach is called the Classical Strategy. In this approach, the pathogen is released, and the population of pathogens slowly increases. The result is a subsequent reduction in the population of the weed.

In only a few instances has the classical approach been successful. The most useful fungi have been rusts. This is due to their capacity to form specific associations with their host, they are capable of air-borne dissemination and they can cause epidemics in ideal conditions.

The rust fungus Puccinia chondrillina has been successfully established on Skeleton weed (Chondrilla juncea), formerly a major weed of wheat throughout the eastern states. The weed has been reduced to tolerable densities. Interestingly, the initial development of control was associated with a change in the genetic structure of the remaining plant population, with weed genotypes tolerant of the fungus surviving. Release of rust fungi for control of blackberries, Parthenium grass, Mimosa pigra and Rubber vine (Cryptostegia) has also been moderately successful.

Use of natural pathogen populations to reduce populations of weeds has a number of problems, including:

Problems

The pathogen must be highly specific and the host highly susceptible. The pathogen must be sustained at levels that enable control of weeds should conditions change to favour the weeds. The pathogen must be able to mutate or change genetically to infect new races of weed. Other control measures should be available (chemicals, insects, mechanical) should the usual variation in environment result in an explosion in the size of the weed population and suppression of the pathogen. Lack of a commercial incentive for organizations to find new biocontrol agents.

Inundative Application of Pathogens

In recent decades, the search for fungi that can control weeds when applied in an inundative manner (fungal products are also known as Bioherbicides) has led to several commercially attractive products.

Effective bioherbicides must have:

1. Specificity to the weed and no other plant.
2. Virulence and pathogenicity to kill the weed when it is among its crop companion.
3. Reproduction in artificial conditions, forming stable and long-lived reproductive units.
4. Capacity to be delivered to control the weed.

Note that this process aims to kill the weed, much in the way a chemical herbicide functions. The pathogen does not survive in the crop after application. The control agent is just another commercial product.

The fungus Colletotrichum orbiculare has been developed to control Bathurst bur (Xanthium spinosum). The weed forms spiny seed casings that can damage the feet and contaminate wool of sheep. The pathogen was first discovered in the early 1980s. The development of the bioherbicide took over 10 years. One major stumbling block was the close relationship between the weed and an important crop species, sunflower. However, specific races were found, and subsequently used to establish the effectiveness of control. Spores of the fungus are suspended in vegetable oil, sprayed on young seedlings and considerable damage develops with 2 weeks.

Constraints to greater use of mycoherbicides include:

Biological: Genetic variation in host, host resistance to infection, interaction with other microbes on the phylloplane, host range, genetic stability of the fungus.

Environmental: Germination requirements, especially temperature and moisture, spore formation and survival.

Commercial: Small returns.

PLANT PATHOGEN CONTROL

Plant disease is caused by abiotic and biotic factors. While other chapters deal with the way fungi improve plant health directly, this chapter will ask the question "Can fungi be used to reduce the effects of biotic causes of plant disease?" As most biotic causes are fungi, we will concentrate on fungus to fungus interactions. The topic of fungus to fungus interactions has been reviewed in depth on many occasions from many different points of view.

The interactions between fungi can be manipulated to reduce the damage caused by one fungus on the plant. Essentially, these interactions include:

1. Direct hyphal parasitism.
2. Antibiotic interactions.
3. Competition for resources.

The mechanisms are not mutually exclusive and all may be evident within one interaction at the same or different times.

Direct Hyphal Parasitism

Mycoparasitism is where one fungus derives its nutrition from another without any benefit in return. The interaction can be where the parasite is biotrophic or necrotrophic. Biotrophs tend to have highly specialised relations with specific hosts. The parasite is extremely difficult to manipulate, and as a consequence few have ever been used in biocontrol.

Necrotrophs tend to kill the cell or tissue from which they feed. They tend to have a broad host range, and their activity commonly includes excretion of antibiotics. For instance, the necrotroph Trichoderma has the capacity to parasitise a great number of fungi, including pathogens Rhizoctonia, Scerotinia, Fusarium and Verticillium, and the Straminopiles Pythium and Phytophthora. Trichoderma grows towards hyphae of other fungi where it binds following branching. Trichoderma releases hydrolytic enzymes that digest the walls of the host prior to penetration. Trichoderma is also thought to release a range of toxins that reduce any response from the host to invasion.

For biocontrol to be successful, the parasite must be placed in an environment favourable for its competitive success. Most soils are oligotrophic. Thus, the introduced fungus is unlikely to be favoured against isolates that are adapted to the local conditions. By judicious increases of organic amendments at inoculation, biocontrol can be attained followed by a decline in the population size of the parasite as the amendment degrades.

Antibiotic Interactions

Antibiosis can be defined as the inhibition of one organism by metabolites of another. Metabolites that kill another organism tend to be called toxins. Thus, dilute solutions of toxins may inhibit and high concentrations of antibiotics may kill.

Most fungi produce inhibitory metabolites. For instance, Gliocladium produces a diketopiperazine that kills Pythium probably because of coagulation of proteins in the cytoplasm. Use of the fungus on seed appears to reduce seedling damping-off as much as a common seed treatment. Volatile pyrones produced by Trichoderma appear to reduce damping off caused by Rhizoctonia. The potential to exploit these metabolites in biocontrol appears feasible.

The role of antibiotics in biocontrol is still unclear. Some spectacular experiments, when repeated in the field, have resulted in enormously variable outcomes indicating that other mechanisms may be important.

Competition for Resources

Competition is the active requirement for resources in excess of those immediately available to two or more organisms. The resource most commonly considered to be important for fungi is organic energy. Thus much competition exists for access to energy for growth and maintenance.

However, competition between fungi is extremely difficult to quantify. The mycelium is constantly changing. Resources within the hyphae are constantly being shifted between different parts, with the consequence that over very short periods of time the nature of what is being assessed changes.

However, competition for nutrients can take place between survival propagules, germinating units and mycelium in any arrangement. The result of competition is stasis of the less successful competitor. Inoculum density does not decline and the potential to initiate disease remains. Thus the application of competition by itself to biocontrol appears limited.

Some fungi used against fungal pathogens

Agent	Pathogen
Chaetomium	Venturia
Coniothyrium	Scerotinia
Cladosporium	Botrytis
Gliocladium	Rhizoctonia
Trichoderma, Penicillium	Rhizoctonia
Scerotinia, Fusarium, Verticilium	Puccinia

Trichoderma viride, Trichoderma harzianum, Trichodermapolysporum and Trichoderma hamatum (all Deuteromycota) have all been formulated for commercial use against soil fungi. These fungi are common inhabitants of heated soil where the almost sterile medium is conducive to their explosive colonisation following germination of ascospores of the fungi. Over time other fungi colonise the soil, competitively excluding the species of Trichoderma. Thus their obvious presence in soil is short-lived. They colonise the soil, interact with the pathogen but then generally decline. For instance, Trichoderma can be used in potting media to reduce seedling diseases, but the fungus is unlikely to be successful in field soils.

Conclusion

While laboratory and glasshouse experiments indicate fungi have a role in biocontrol of plant pathogenic fungi, the results of field application of a single fungus have been discouraging. While it appears that some agents may have a place in integrated approaches to reducing plant pathology, we do not appear to understand the basis of interactions between fungi. Until this understanding is improved, use of fungi in biocontrol is likely to remain a dream.

NEMATODE CONTROL

Fungi utilise many different sources of carbon, some living and some dead. The observation that nematodes can be colonised by fungi, led to the possibility that the interaction might be used to control pest nematodes of plants and animals. Closer examination of the interaction has led to the belief that only partial control of nematodes is possible and that in a practical sense, populations of pests may be suppressed not eliminated. In the case of plant and animal parasitic nematodes, fungi may be used to regulate nematode populations in conjunction with other processes.

Pest Nematodes

Nematodes found in plants, animals and soil use various sources of organic energy. Plant parasites feed from plant cells, either by selecting and then establishing a single feeding site. This approach is usually known as sedentary feeding. Other plant-parasitic nematodes are migratory, moving from site to site on the root, rarely using a single cell for a long time. Migratory plant parasites can be further categorised as internal and external feeders, meaning that they migrate within or outside the root.

Animal parasites of agriculturally important ruminants tend to have a life cycle whereby colonisation of the host is followed by a soil phase. Larvae are consumed by grazing animals along with grass. The larvae establish a feeding location within the digestive system of the host. Adults lay eggs within the intestines, and the eggs fall to the ground with the faeces. Eggs subsequently hatch in the faeces and larvae move to blades of grass, where they are consumed.

Fungi that colonise nematodes fall into several different categories, each of which requires a different approach if it is to be used successfully to control pest nematodes. The fungi range from highly specialised nematode parasites, to generalists with no specific host. Some fungi may colonise organic matter and incidentally also colonise eggs, larvae or adult nematodes. Other fungi may trap larvae or adults and colonise their body, using the body as the primary source of nutritients.

Plant Parasitic Nematodes

Endoparasitic fungi are generally highly specific to single species or genera of nematodes, and are difficult to culture in the absence of their host. Hirsutella, Meria, Nematophthora and Nematoctonus have all been suggested as ideal biocontrol agents because of the specificity for economically important nematodes. However, the difficulty of culturing them and lack of experimental support for their activity has limited any commercial development.

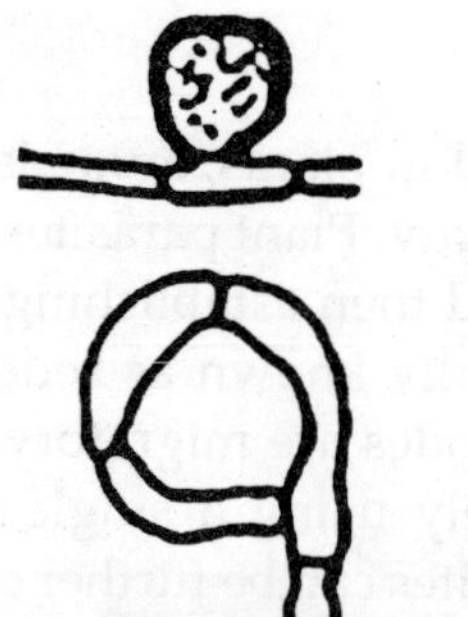

Fig. Two basic types of trap found in fungi, the sticky nob or the constricting loop.

Nematode trapping fungi utilise either sticky pads or constricting rings to immobilise nematodes. The fungus then digests the internal organs of the nematode and form reproductive cells. The main fungi in this group are Arthrobotrys and Duddingtonia. These fungi are found in soil where they live on readily-available organic carbon, in the absence of nematodes. They are generalist feeders, being able to trap and digest many nematodes, not just those that infest plants. Thus, the fungi are rare, unable to exist at sufficiently high densities to control nematode populations, and parasitic on beneficial as well as pest species of nematode. Because of these difficulties, exploration of the use of trapping fungi in the field has largely ceased.

Parasitic fungi have been isolated from either eggs, larvae or adults nematodes. Several of these have been found to preferentially parasitise the nematode and thereby reduce the size of the nematode population. Research efforts in some crops are now concentrating on using fungi found locally to control local pests. The general approach is to go to locations where nematodes have reached high densities. Parasitised individuals are extracted from soil, the fungi cultured and then tested as parasites of the pest nematode. The theoretical basis for this approach is that high densities of nematode will enable fungal parasites to increase population size dramatically. Thus by selecting fungi adapted to the host (nematode) and environment, effective and usually inundative inoculation will enable the control of the pest nematode. Biocontrol in problem fields has been achieved in a surprising number of cases. The species used include Paecilomyces lilacinus, a fungus that appears to have a wide range of potential hosts, both insect and nematode, and yet isolates have a degree of specificity for host.

Animal Parasitic Nematodes

Sheep and cattle have traditionally been drenched to control parasitic nematodes. However, nematodes are now developing resistance to the common active ingredients in drenches. Drenches using different active compounds are not being developed because the economical returns are low for the investment. Thus alternative approaches are being sought.

In one approach, nematode trapping fungi are placed into nutrient-rich faeces. As nematodes hatch, they become colonised by the fungus. The fungus Duddingtonia has been inoculated to the dung via a drench to the animal. Sclerotia of the fungus pass through the intestine, hatch in moist dung, and effectively reduce larval densities IN SOME CASES. While further research is needed before commercial adaptation.

INSECT CONTROL

Most people who garden are aware of insects that have been apparently colonised by microbes. The small discoloured carcasses of larvae or adults are unearthed as you dig through soil. One group of organisms that parasitise insects are the fungi. Insects are common pests of agriculture and horticulture. Several fungi, like those seen in parasitised garden insects, have been used to control insect pests.

A diverse array of fungi associate with insects, and the nature of the interaction is also diverse. Biocontrol of insect pests can take one of several different approaches. The most common approaches are inundative application and classical control.

Fungi Used Inundatively

Several species of fungus have a reputation for biocontrol of insects. The fungi come from all divisions. The most commonly used ascomycetous fungi include the genera *Beauvaria,* Metarrhizium, Tolypocladium and Verticillium. These fungi have been applied inundatively, in the manner of a chemical application, with the goal of killing most, if not all, insects.

The most historically important of the commonly used fungi is Beauvaria bassiana. The Chinese have a long history of successfully culturing silkworms. However, when the industry was taken to Europe, cultures of insects were devastated by Muscadine disease. The causative agent in turn was determined as Beauvaria bassiana, a fungus that grew on and in larvae. Determination by Bassi that the infectious agent was a fungus, where germs were shown to cause disease. The fungus has since been used successfully to control the larvae of many Lepidopteran insects.

Metarrhizium, also an asexual stage of an Ascomycete, has been successfully used to control spittlebugs in South

America, Rhinoceros beetle in the South Pacific islands, and in Australia to control locusts. Lecanicillium lecanii infects sucking insects such as aphids and scale insects. Some plant-sucking aphids also have the nasty habit of transmitting viruses, thus control of the insect and the viruses it transmits are both possible. The fungus is used to control sucking insects mostly in greenhouses where humidity is high. Tolypocladium is a fungus found in both soil and water. It has been reported to colonise a wide range of insects, and effectively controls mosquitos in experimental conditions.

Mode of Attack

Penetration: In general, when a spore comes in contact with the insect cuticle, the spore attaches, germinates, penetrates the cuticle (from outside in, or inside out, depending on whether the spore lands on the outer surface, or is ingested). Penetration may involve physical force, or more commonly physical and enzymic action combined. Inside the haemocoel, the form of the fungus may change to yeast-like, though filaments and wall-less protoplasts are also found.

Secondary Metabolites: Many insect pathogens release metabolites that increase the chance of insect death. A wide range of pathogenicity factors are known and they come from different metabolic pathways. The factors may be toxins, immunosuppressants or antibiotics. Destruxens from Metarrhizium are polypeptides that appear to be directly toxic to the insect. Cyclosporin-like immunosuppressants from Tolypocladium increase the rate at which the fungus colonises the haemocoel. Better understanding of the formation and mode of action of secondary metabolites will be important if rapid knock-down of insects remains the goal of inundative biocontrol.

Other Approaches to Insect Control

The control of pasture cockchafers has been achieved in the Adelaide Hills of South Australia by the release of an isolate of Metarrhizium specific to the host. The larvae of the insect burrow through soil where they feed from roots of pasture grasses and incidentally, contact spores of the fungus. Colonisation of the larvae results in death after several days, with the subsequent release of spores of the pathogen, followed by subsequent cycle of contact, colonisation and insect death.

In principle, the density of the fungus increases when high levels of insect are present in the soil. As a result the insect population declines, enough for most of the larvae to miss contact with the fungus. However, the fungus is sustained at a low density, to become active only as the larval densities increase. The classic pattern of population change is shown in the figure.

Variations

Several insects are of major concern in Australia. One, Heliothis (Helicoverpa) armigera is a serious pest of many different crops and pastures. In cotton, larvae eat foliage and bore into boles. Heliothis has several stages where it is vulnerable to biocontrol agents: as eggs on the plant surface, as larvae on the plant surface and as long-lived pupae in the soil. Fungi may play a role at each stage. Fungi that live on the leaf surface may colonise eggs. Fungi that live in the leaf may deter or poison larvae feeding from the leaf, or be treated using inundative approaches, and fungi may colonise pupae in soil.

MUSHROOM PRODUCTION

A huge range of edible fungi are cultured. Agaricius bisporus is cultivated widely in western countries. In addition, Shii-take (Lentinus edodes), Straw mushroom (Flammulina velutipes), Oyster mushroom (Pleurotus ostreatus), and Chinese black mushroom (Auricularia polytricha) are cultured in various Asian countries, and increasingly in western countries. The truffle industry is economically important in southern Europe and production is expanding in the southern hemisphere. Finally, edible mushrooms are picked from the wild, especially in Europe. Many of the fungi are eaten fresh, but there is also a market for dried mushrooms and canned truffles.

Button Mushroom

The identity of the fungi grown as commercial or button mushrooms in western countries is not altogether clear. The generally accepted name is Agaricus bisporus. However, breeding over the years appears to have resulted in the infusion of some genetic material from various sources.

The commercial mushroom is grown in a process that varies only slightly around the world. Essentially, straw is mixed with stable and poultry litter, mixed with appropriate minerals and then composted. Spawn is added after the compost has cooled after the second phase. The compost is then laid out, usually in purpose built sheds in which the temperature is held at around 24 degrees. A layer of inert, alkaline material is added as a casing over the surface. Once mycelia reach the surface of the beds, the temperature of the shed is reduced to induce fruiting. Fruiting then proceeds over the next few weeks. Commercial mushroom production is a highly intensive industry. Understandably, pests and diseases are a major problem. The principles of control do not differ from other intensive production systems.

Shii-take and Log Culture

Shii-take is a highly prized mushroom in Japan. The range of variation in flavours and odours make this a valuable ingredient in the usually delicately flavoured Japanese recipes. The range comes from the genotype of the spawn, selection of logs used to grow the fungi, and conditions under which the logs are then incubated. Shii-take is commonly grown on logs of deciduous trees. The logs are inoculated with spore suspension placed over the logs or by overgrowth from pre-inoculated fragments of timber. The base logs are not sterilised so the process has many potential hazards.

After extensive period of colonisation of the logs (laying), the induction of fruiting takes place when the logs are stood upright and the temperature drops. After a period of maturation, the temperature increases. Fruiting then takes place in spring, and continues for some years after.

Cropping takes place in the open, and traditional approaches continue to be used in Japan. Most of the crop is sold fresh, though some is dried, packaged and then sold throughout the year. The dried crop is most commonly exported from Japan. Artificial approaches to cropping have been developed. These use controlled temperatures, and selection of particular strains. In Australia, cropping more commonly uses these artificial environments and selected strains.

Organic Waste

Straw mushrooms are traditionally grown on straw from the rice paddy. The bundles of straw are drenched and then stood upright. The bundles are usually inoculated with spawn. Each bundle can produce fresh mushrooms for several weeks. More recently, bundles of straw have been mixed with cotton or other organic waste, or cotton waste used exclusively. These alternatives often produce more mushrooms, probably due to the improved N:C ratio.

Ectomycorrhizal Fungi

Fruiting bodies of a number of ectomycorrhizal fungi are consumed. The Black truffle (Tuber melanosporum) is perhaps the most highly prized of these fungi. Other truffle species of Tuber, Terfezia and Tirmania are also eaten. Field collections of morels (Morchella sp) and chantarelles (Cantharellus cibarius) are also highly prized. The latter fungi are found naturally in Australia, and sometimes appear in the shops in autumn.

Truffles have been harvested in European countries for many centuries. True Black truffles are found mainly in France, and other species of Tuber in France, Spain, Italy and many countries surrounding the Mediterranean sea. The Black truffle is the most highly prized, because of its very strong flavour. Culture of Black truffles is indirect, in that trees that support the fungi are managed with the aim of maximising the production of the fruit bodies. The soil of the truffiere tends to be alkaline, calcareous, and well drained. Oak or Hazel trees are inoculated in the nursery. Saplings are plants out in the truffiere. The trees are then pruned to encourage lateral growth. After about seven years, the fruit body appears in late winter to spring. The vegetation above the zone of fruiting tends to die off. Pigs or dogs have been used to detect ripe truffles, though more biochemical methods are being developed. The truffiere continues to produce abundantly for around 30 years.

DISEASES OF FUNGI

When any production system uses an organism of a single genotype under highly intensive conditions, then the chances of a mutation in an associated microbe or virus causing an outbreak of disease increases dramatically. Viruses have a small genome. Increased virulence resulting from a chance mutation is likely.

The viruses that have been detected in fungi are all thought to be double stranded RNA. A protein coat may envelope the virion (nucleic acid). Further, some encapsulated virions are wrapped in a membrane of host origin. The observation of a single structural type must be viewed in recognition that current searches for viruses in fungi assume a structure consisting of double stranded RNA. Other forms may be ignored. Further, some potential viruses might also be considered as transmissible genetic elements, and given a different name. Care must be taken when reading the literature on fungal viruses.

Viruses are thought to be ubiquitous in fungi. Estimates of their presence suggest that 30 per cent of species may be colonised. Most viruses have not been associated with any form of noticeable pathology, or expression of phenotypic change, and are thought to be benign. All viruses have been found within fungi and their existence outside the fungal host is unknown. The lack of an extracellular phase has implications for transmission.

Transmission

In those few associations where transmission has been studied, the virus is transmitted vertically or horizontally, or both. Vertical transmission refers to the movement of virions into the reproductive units of the fungus, to then be expressed in the next generation. The virions may be carried in the spores. Following germination of the spore, the virus appears in the next generation.

Horizontal transmission is where fusion between compatible hyphae results in the movement of cytoplasm containing the virions to the uninfested mycelium. The importance of vegetative compatibility cannot be overemphasised. Compatibility determines the potential for transmission, and is thus enormously important in the ecology of the virus, especially when infection is associated with disease.

La France Disease of Mushrooms

Cultivated mushrooms have been examined for viruses especially after reductions in productivity. Although several different viruses appear to be pathogens, La France disease has been most widely studied.

The presence of an isometric particle has been consistently associated with the disease. Infected cultures grow more slowly. Basidiocarps of infected tissue appear earlier, they are smaller and commonly deformed. Total spore production is reduced though spores appear sooner. The disease is highly infectious, and rapid spread through a production system, and between adjacent farms, is possible. The virus can be spread vertically and horizontally. Vertical spread is associated with the early appearance and release of spores from infected mushrooms. These spores anastomose with uncontaminated hyphae and transfer the virus. The virus multiplies rapidly and spread through the contiguous mycelia.

Horizontal transmission means that only a small amount of contaminated mycelia mixed in with spawn may result in the entire bed becoming infected rapidly. Hyphae grow rapidly and establish anastomoses with compatible hyphae (most likely all the bed). Transmission between farms is possible if they are using fungi with compatible hyphae. This is the case when farmers are using a single variety. Use of different varieties reduces the chance of transmission. However, even unrelated varieties may have a low level of compatibility, thereby enabling the occasional anastomosis and transmission of the virus. Complete incompatibility is uncommon, so virus transmission remains a potential problem for all mushroom farms.

Killer Yeasts

Killer Yeasts are strains that are infected with one of a group of viruses that produce a toxic protein that kills closely related but uninfected yeasts. The yeasts belong to a wide diversity of fungi, including the bread-making and alcohol-producing Saccharomyces cereviseae and the maize smut Ustilago maydis. Killer yeasts are tolerant of the toxins they produce, but may be susceptible to toxins of different Killer yeasts. The genes that confer immunity and toxin production appear to be on the virus genome. However, replication and other functional genes of the virus may be on the host nucleus.

The fitness of Killer yeasts in natural habitats is partly attributed to their toxin production. The toxin apparently regulates competition and enables the host cell to proliferate. Killer yeasts are commonly found on fruits and flowers. They are commonly found widely on grapes. They are important for fermentation of grapes, because they may slow fermentation and leave undesirable flavours in the wine. Killer yeasts can take over fermentation by killing the inoculated yeasts, even when the concentration of killer cells is low at the beginning of the process. The potential of Killer yeasts is being examined for commercial exploitation. The capacity to add desirable fermentation characters to strains that already reduce competitors in the must is a valuable attribute. Use of killer strains may reduce the need to sterilise must at the beginning of fermentation. The sterilant, sulphur dioxide, induces asthma and sinus conditions in some individuals, and is therefore seen as an undesirable though necessary additive to wine.

Evolution of Pathogenicity

Pathogenic viruses are only found in a few situations where the fungal population is held at artificially high levels by human intervention. This has led to the suggestion that most viruses are naturally benign. Viruses have become pathogens because chance mutations in the small genome occur at a much faster rate than in the host, and selection is for pathogens when host populations are held at high densities. In other words, disease is a function of the attempts by humans to manipulate the environment. However, the presence of Killer yeasts indicates that the interaction between virus and fungus may be more complex. Where the virus provides a beneficial function for the host, then the virus may become incorporated into the host genome. Competitors may be almost entirely replaced in a population. Thus viruses are not one functional grouping. They have attributes determined by their genome, and we will no doubt find further variants as fungi are used more extensively and intensively.

INDUSTRIAL PRODUCTION

Fermentation

If you were to ask someone what role fungi played in industry, it is likely that they would indicate use of yeasts in brewing, wine making and bread making. This response is probably related to their level of awareness of yeasts in brewing and baking, rather than their understanding of the importance of fungi in industry. Brewing and baking rely upon a simple principle. The basis of current industries has been around for many centuries. Fermentation commonly results in alcohol, but lactate or glycerol can also result. The principles are simple and well understood.

Production of Alcohol

In the yeast Saccharomyces cerevisiae presence of excess glucose represses respiration. In principle, materials rich in sugars (or starches) are then fermented resulting in the production of alcohol.

Alcohol is produced almost invariably by use of Saccharomyces cerevisiae or its close relatives. Ales and wine use S. cerevisiae, lager uses S. carlsbergensis, cider uses S. uvarum, and saki uses S. sake. The yeast hydrolises sugars, commonly sucrose, to pyruvic acid via the Embden-Myerhof-Parnas pathway, and then via acetaldehyde to ethanol. The reaction is exothermic, and unless the heat is dispersed, the reaction is slowed. Further, though up to 50 per cent of sugar can yield alcohol, by weight, the solution rarely goes beyond 15 per cent ethanol, because the fungus is sensitive to high concentrations of ethanol.

Chardonnay, Reisling, Shiraz, Port, Colombard, Grenache, Pinot Noir, Cabernet, Champagne

Approximately half the wine of the world is produced in Italy, France and Spain. Australia and USA are rapidly increasing production of high quality wines. The main grape used is *Vitus vinifera*. As the grape berry ripens, the concentration of sugar in the juice increases and acidity (especially malic acid) decreases. After harvest, the grapes are crushed and the juice (must) is either fermented in contact with skins (red wine from dark berries) or without (white wine).

Fresh berries are covered by various microbes, including yeasts. The contaminating microbes are suppressed by addition to the must of sulphur dioxide, or a compound that releases sulphur. Alternatively, the must might be sterilised by other means. The starter culture of specifically selected yeasts is then added. The temperature is carefully controlled, and various treatments are used to influence the flavours that develop. After a period, the yeasts sink and the development of complexity allowed by aging and maturation.

Beer, Ale, Lager, Stout

Ales and lager use cereal grains, commonly barley, as the basis of fermentation. Grain is dried, then germinated synchronously. The germination process involves release of amylases within the grain. These enzymes are then used to digest the starch of a batch of ground cereals mixed into a mash, resulting in the release of hexoses. The resulting wort is drained off. After various treatments, including addition of hops, the liquid is inoculated with specific yeasts and the mixture fermented for a short period.

Sake

Sake is a Japanese beverage made from rice. Steamed rice is inoculated with Aspergillus oryzae to produce koji. Further rice mash is lactic acid fermented using bacteria and yeasts. The mash and koji are mixed and fermented for around 20 days, when the alcohol concentration has reached around 18 per cent. The product is filtered, pasteurised and stored before consumption.

Bread Making

Bread consists of a mixture of flour (usually from cereals especially wheat), with water, salt and sugar, leavened by yeast. Flour is mixed with the remaining ingredients and incubated at around 25 degrees. The yeast ferments the sugar forming carbon dioxide and alcohol. The released gas causes bubbles to form by elastic extension of gluten (a protein) in the flour. On baking, the alcohol evaporates. The length of leavening, the quantity of gluten in the flour, the constituents of the grain, and the temperature determine the texture of the bread. Flour from wheat can be mixed with starch from a variety of sources. For instance, potato and pumpkin have been used successfully mixed with wheat flour to make a different style and flavoured bread.

DRUGS FROM FUNGI

At the beginning of the 21st century, Fungi were involved in the industrial processing of more than 10 of the 20 most profitable products used in medicine. Three anti-cholesterol statins, the antibiotic penicillin and the immunosuppressant cyclosporin A are among the top 10. Each of these has a turn over in excess of $1 billion annually. Fungi are extremely useful organisms in biotechnology. Fungi construct unique complex molecules using established metabolic pathways. Different taxa produce sets of related molecules, each with slightly different final products. Metabolites formed along the metabolic pathway may also e biologically active. In addition, the final compounds are often released into the environment. Manipulation of the genome, and environmental conditions during formation of compounds, enable the optimisation of product formation.

On the negative side, single isolates of fungi in manufacture may lose their capacity to form or release the target molecules. Indeed, the target compound may only be expressed under specific conditions, or at a specific point in the life cycle of the fungus. It is amazing that so many biologically active compounds have been discovered and taken to point where they are medically important. The role of fungi was established early in history. Yeasts have been used in the making of bread and alcohol since the beginning of civilisation. In modern times, the discovery of penicillin marked the beginning of a new approach to human health and established the importance of fungi.

Antibiotics From Fungi

In 1941, penicillin from the fungus Penicillium chrysogenum was first used successfully to treat an infection caused by a bacterium. In doing so, it revolutionised treatment of disease. Many formally fatal diseases caused by bacteria became treatable, and new forms of medical intervention were possible.

When penicillin was first produced, the concentration of active ingredient was approximately 1 microgram per ml of broth solution. Today, improved strains and highly developed fermentation technologies produce more than 700 micrograms per ml of active ingredient. In the early broths, several closely related molecules were present. These molecules are beta lactam rings fused to five-membered thiazolidine rings, with a side chain. The side chain can be chemically modified to provide slightly different properties to the compound.

The natural penicillins have a number of disadvantages. They are destroyed in the acid stomach, and so cannot be used orally. They are sensitive to beta lactamases, which are produced by resistant bacteria, thus reducing their effectiveness. They also only act on gram positive bacteria. Modifications to manufacturing conditions have resulted in the development of oral forms. Antibiotic resistance among bacteria is becoming an extremely important aspect determining the long-term use of antibiotics.

Cephalosporins also contain the beta lactam ring. The original fungus found to produce the compounds was a Cephalosporium, hence the name. As with penicillin, the cephalosporin antibiotics have a number of disadvantages. Industrial modification of the active ingredients has reduced these problems. The only useful antifungal agent from fungi is griseofulvin. The original source was Penicillium griseofulvin. Griseofulvin is fungistatic, rather than fungicidal. It is used for the treatment of dermatophytes, as it accumulates in the hair and skin following topical application. The antifungal agents are readily and cheaply produced industrially.

Immune Suppressants

Cyclosporin A is a primary metabolite of several fungi, including Trichoderma polysporum and Cylindrocarpon lucidum. Cyclosporin A has proven to be a powerful immunosuppressant in mammals, being widely used during and after bone marrow and organ transplants in humans. Cyclosporin A is a cyclic peptide consisting of 11 mainly hydrophobic amino acids. Its inhibition of lymphocytes was first discovered during the 1970s. Subsequently, the mode of action was elucidated.

Cyclosporin A binds to a cytosolic protein called cyclophilin. Cyclophilin is found amongst many different organisms and its form appears highly conserved. Cyclophilin is involved with folding the protein ribonuclease. However, the Cyclosporin A/cyclophilin complex also binds to calcineurin. Calcineurin dephosphorylates a transcription factor, thereby triggering transcription of numerous genes associated with T cell proliferation. When the complex binds to calcineurin, T cell proliferation is suppressed. The inhibition of T cells proliferation results in the suppression of the activation process associated with invasion by foreign bodies. As a consequence, transplant tissues, which are foreign bodies, are not rejected.

Calcineurin is also highly conserved amongst phylogenetically diverse organisms. In fungi such as the human pathogen Cryptococcus neoformans, calcineurin is necessary for recovery from cell cycle arrest, growth in hypertonic solutions and regulation of the calcium pump. Thus the interaction of the Cyclosporin A/cyclophilin complex with calcineurin in Cryptococcus will result in death of the pathogen. However, in humans, cyclosporin also suppresses the immune system. The side effect is an unacceptable risk, and Cyclosporin A is not used as a fungicide in humans at present.

Gliotoxins also have immunological and antibiotic activity. Produced by many fungi including Aspergillus fumigatus, gliotoxins belong to a class of compounds called epipolythiodioxopiperazines. The antibiotic activity is widely recognised and considered uninteresting. However, its effect on the immune system, especially macrophages, is being re-examined. A wide range of other compounds with antibiotic activity are also known. They have been rejected for use in medicine because of unwanted side effects, or instability of the active compound.

Ergot Alkaloids

Claviceps purpurea is the causal agent of St Anthonies fire, a scourge of the middle ages when ergots contaminated flour. The ergots contain many alkaloids. Their effects are quite variable. They act on the sympathetic nervous system resulting in the inhibition of noradrenaline and sclerotin, causing dilation of blood vessels. They also act directly on the smooth muscles of the uterus causing contractions, thus their early use to induce abortion. Their strongest effect is intoxication, caused by lysergic acid amides, one of which is the recreational (and illegal) drug, LSD.

Ergot alkaloids have a number of medicinal uses. Perhaps the most widespread use is in the treatment of migraines. The vasodilator activity reduces tension during an attack. The drugs also reduce blood pressure, though with untoward side effects. Alkaloids are now produced in culture by strains of C. fusiformis and C. paspalii.

Statins

Aspergillus griseus, a soil-borne fungus, produces secondary metabolites that have a different range of effects. The molecules called "STATINS" have been used to reduce or remove low density lipoproteins from blood vessels in humans. In fact, the compounds all act via an enzyme in the liver that makes cholesterol. By blocking the enzyme, the body removes cholesterol complexes from the inside of blood vessels. This has the effect of reducing or removing blockages in arteries, and thereby reducing the chance of a heart attack, strokes and diabetes.

In addition, statins have been implicated in attracting stem cells to damaged tissues. The stem cells then appear to regenerate the tissue. Some statins induce problems. One form of the drug has been associated with muscle wastage. Others appear to lack side effects and have been recommended for wide spread use to control heart disease.

FOOD PROCESSING

Fungi may be the food, or may make the food edible following processing. Processing may make it possible to consume the foodstuff by adding, modifying or removing components, including flavours, nutritional elements such as vitamins or colours to enhance the appeal of the food.

Fungi are a common contributor to the processing of foods. Their use dates back to the start of the civilisation, when breads and wines were first made deliberately. These days, the selection and use of fungi is a highly organised field of research and development.

Soy Sauce

We tend to take a variety of foods and food additives for granted, without being aware of the processes which get them to the table. One such food is Soy Sauce and its partner bean curd.

Soy sauce (shoyu) is a dark brown, salty liquid, high in amino acids and with a meat-like flavour. It was first produced in Japan (a similar product is made in other east Asian countries), where some microbial cultures were used to ferment the unpalatable soy beans. The current industrial process is highly controlled, used around the world and is based on this original process.

Fermentation is in two stages. Initially, soy beans are soaked, cooked to remove contaminants, and then mixed with roasted wheat. A culture of the fungus Aspergillus oryzae is added to the mix and kept aerobically for 20 to 40 hours at 25 degrees. The fungus produces invertases, amylases and cellulases, which degrade the soy paste. The paste is then mixed and taken into the second phase of fermentation.

In deep vats, brine is added to the paste and the yeast Saccharomyces rouxii and lactobacilli are added. Anaerobic conditions develop quickly, preventing further growth of A. oryzae. After about a month, a sour liquid is apparent. The liquid contains large concentrations of amino acids, simple sugars and a range of vitamins. After separation and further storage, the liquid is sterilised, bottled and sold as Soy Sauce. Similar products are called Koji, Idli, Patu, Laochao or Ogi.

Blue Cheese

Various cheeses are sold that have been stab-inoculated with a strain of Penicillium roquefortii. The result is a blue streak or vein through the cheese. The fungus imparts a strong, pungent flavour due to the aerobic production of methyl ketones. Famous blue cheeses include: Roquefort, Gorgonzola, Stilton, and Danish Blue.

Other Flavours

While many flavours are produced by bacteria, fungi are responsible for a range of flavours including terpenes, menthol and lactones. Fungi also produce compounds that deodorise offensive and neutralise bitter flavours. At present, flavour enhancement is an unimportant area of industrial use of fungi.

Colours

Fungi produce a range of compounds that alter the colour of food. For instance, Monoascus purpureus has been traditionally used for the production of red wine. The pigments are polyketides that are insoluble in acid conditions. Beta carotene is produced by a range of Mucorales. This can be added to a variety of foods. Concern with the potentially toxic or allergic characteristics of some artificial colours has led to a closer examination of colours from natural sources.

PAPER MAKING

Art Papers From Fungi by Allein Stanley

The basic process of making paper is straightforward and no different for mushrooms than for other fibres. The one significant difference is that you are using a chitin base rather than a cellulose base for the paper. There are several variations on the theme. The following are very flexible guidelines, and should be adapted to your specific situation.

Deckle and Mould

The deckle is a frame the same size as the mould. Both it and the mould can be made with wood strips shaped into a rectangle, like two 'L' shapes put together. The screen will be stapled or tacked to the top of the mould.

The deckle is placed over the mould when picking up your fibres and it also 'edges' your paper. If you wish to simplify, you can dispense with the deckle and use screen wire with all four sides covered with duct tape to use as a mould. In this case, it is very important to hold it flat while lifting underneath the slurry. The edge may not be as even but that may be a desirable characteristic.

Preparation and Process

Make a thick stack of newspapers, over which place towelling or other absorbent materials.

Making the Stock

After soaking, chop the mushrooms and grind with water in a blender for a minute or two. Blend on different settings until we get a puree. Do not overload the blender, repeat blend instead. If you wish to add bits of soaked paper or rags strips, add them at this time.

Slurry

Dump the stock into the tray with plenty of water. The amount of stock to water is something you will need to trial. Stir until materials are well distributed. With your hand, move from side to side under the surface of the water to more or less line up the fibres. Holding the 'deckle and mould' on each side, tilt it under the surface and quickly lift it up underneath the floating materials. Very quickly tilt it in each direction to get good coverage. If the screen is not well covered, dip it again until you are satisfied. Allow the paper to drip back into the tray until most of the water has run off.

Couching

Place your hands so that thumbs are underneath, fingers on top of each side of the mould and quickly flip paper onto the sheeting or towelling. Screen should now be on top of your paper. With sponge, sop up excess water. Carefully lift the screen from your paper. If it has become too dry, the paper will tear or the screen will stick to it. If this happens, sponge on more water through the screen until you can get it to release. The thinner the paper, the more delicate this step becomes.

Drying

Continue to dry by replacing newpapers and covering cloths. Ironing gently over a cover cloth can speed up this process. When dry enough to be easily handled, hang to complete or, if you wish very flat paper, put it under weights, continuing to change underneath papers or cloths frequently.

11

Treatment of Fungal Infections

MYCOLOGY FUNDAMENTALS

Fungi can appear microscopically as either rounded, budding forms (yeastlike organisms) or hyphae. Yeastlike colonies are smooth, while mold colonies are fuzzy; fungi that grow as yeasts include species of *Candida* and *Cryptococcus*, while fungi that grow as molds include species of *Aspergillus*, *Rhizopus*, and dermatophytes (ringworm fungi). The fungi that cause histoplasmosis, blastomycosis, sporotrichosis, coccidioidomycosis, and paracoccidioidomycosis are called *dimorphic* ("having two forms") because they are spherical in tissue but grow like molds when cultured at room temperature. *Candida* species other than *Candida glabrata* appear in tissue as both budding yeasts and tubular elements called *pseudohyphae*. *Pneumocystis carinii* is closer to fungi than to parasites by ribosomal sequences. Because the drugs used to treat *Pneumocystis* pneumonia are also used to treat parasitic or bacterial infections.

Many fungi can form two different types of spores and are given different names, depending on the spore-bearing structures. When the spores are produced by mitosis, the fungus is said to be an *anamorph*, or to be in the imperfect state. Many fungi can have different sporulating structures in which genetic recombination occurs, often as a result of coculture with a strain of the opposite mating type. A fungus producing those distinctive spores is said to be a *teleomorph*, or to be in the perfect state. Diagnostic laboratories usually use the name of the anamorph because they do not use culture conditions that would produce the teleomorph. One exception is *Scedosporium apiospermum*, which is often observed as a teleomorph in the diagnostic laboratory and identified as *Pseudallescheria boydii*.

Most fungi that are pathogenic for humans are saprophytes in nature; they cause infection when airborne spores reach the lung or paranasal sinus or when hyphae or spores are accidentally inoculated into the skin or cornea. Acquisition of infection from another person or an animal has been reported in the case of ringworm but is very rare in other mycoses. Thus, hospitalized patients with fungal infections do not require special isolation. Most fungi infect hosts preferentially by one route and only infrequently by other routes. The agents of ringworm, pityriasis versicolour, and piedra infect the epidermis and its appendages. Sporotrichosis and mycetoma usually arise from subcutaneous inoculation. Inhalation is the route of inoculation for the agents of most deep mycoses. Ingestion of fungi rarely causes infection; *Candida albicans*, a normal commensal in the mouth and intestine, reaches deeper tissues only when mucosal or cutaneous barriers are breached by disease, surgery, trauma, or catheterization. Histoplasmosis, blastomycosis, coccidioidomycosis, and paracoccidioidomycosis have been called "endemic" mycoses to emphasize their restricted geographic distribution. Some fungi, such as *Aspergillus*, are said to be opportunists in that they usually infect hosts with compromised immunity. This distinction is relative, not absolute.

Immunity after exposure to fungi may confer partial protection against reinfection. Residents of areas in which mycoses are endemic are less subject to infection than are newcomers. Predisposing factors are helpful in defining host defence. Immunoglobulin deficiencies do not appear to predispose to any mycosis, whereas neutropenia is common among patients who develop invasive aspergillosis or deep candidiasis. Cell-mediated immunity appears to be of paramount importance in most other deep mycoses.

DIAGNOSIS

Many fungi can be identified to the genus or even the species level by microscopic examination of smears or biopsy specimens. Calcofluor white staining with fluorescence microscopy is a sensitive technique for smears of sputum, bronchoalveolar lavage fluid, or pus. India ink smear remains the method of choice for detecting cryptococci in cerebrospinal fluid (CSF). *Candida* yeast cells and pseudohyphae are the only fungi that are usually gram-positive on smears. For other fungi, Gram's staining is distinctly suboptimal. For histopathology slides, Gomori methenamine silver and a neutral counterstain are preferred.

The method used has a marked effect on the rapidity and sensitivity of blood cultures for fungi except in the case of *Candida* species, which are relatively easy to grow. For most other fungi, concentration of the blood by lysis centrifugation and culture on solid medium constitute the optimal technique. Commercially available nucleic acid hybridization techniques can speed the identification of slow-growing molds, such as *Histoplasma capsulatum* and *Coccidioides immitis*. Serology has limited value, but testing of serum or CSF for cryptococcal antigen or antibody to *C. immitis* can be diagnostic. Detection of *Histoplasma* antigen in urine or serum is helpful in diagnosis and in following the results of treatment for disseminated histoplasmosis. Skin testing with fungal antigens is not useful in detecting active infection.

Antifungal Therapy

Topical Agents

Imidazoles and Triazoles. These synthetic compounds act by inhibiting ergosterol synthesis in the fungal cell wall and, when given topically, may cause direct damage to the fungal cytoplasmic membrane. The imidazoles available for cutaneous application include clotrimazole, econazole, ketoconazole, sulconazole, oxiconazole, and miconazole. Vaginal formulations include four imidazoles (miconazole, clotrimazole, tioconazole, and butoconazole) and one triazole (terconazole). As yet, no substantial differences in the efficacy of or local intolerance to the various topical azoles have become apparent. All are effective in the treatment of cutaneous candidiasis, tinea (pityriasis) versicolour, and mild to moderately severe ringworm of the glabrous skin. Vaginal formulations are effective for vulvovaginal candidiasis. Clotrimazole is poorly absorbed from the gastrointestinal tract, but the oral troche is useful as a topical treatment for oral and esophageal candidiasis.

Polyene Macrolide Antibiotics These broad-spectrum antifungal agents combine with sterol in the fungal cytoplasmic membrane, increasing membrane permeability. Topically, they are not active against ringworm but are effective against candidiasis of the skin and mucous membranes. Nystatin and amphotericin B suspensions are effective in oral thrush, and vaginal troches are effective in vulvovaginal candidiasis. Both nystatin and amphotericin B are available in topical preparations for cutaneous candidiasis.

Other Topical Antifungals Ciclopirox olamine, haloprogin, terbinafine, and naftifine have the same clinical spectrum among the cutaneous mycoses as the imidazoles. Tolnaftate and undecylenic acid are effective against ringworm but not candidiasis. Keratolytic agents, such as salicylic acid, are helpful as accessory drugs for some hyperkeratotic skin lesions.

Systemic Antifungals

Griseofulvin Griseofulvin is a useful drug in the treatment of certain kinds of ringworm; however, it is ineffective in the treatment of candidiasis. The microcrystalline and ultramicrocrystalline preparations differ in dose but not in efficacy. Absorption of both is enhanced when the drug is ingested with fat-containing foods. Griseofulvin interacts with phenobarbital and coumarin-type anticoagulants.

Terbinafine Oral terbinafine (250 mg once daily) is at least as effective as itraconazole and more effective than griseofulvin in onychomycosis and ringworm. Treatment duration ranges from 3 months for fingernails to 6 months for toenails. Gastrointestinal distress is the most common side effect. Rash, hepatitis, and pancytopenia have occurred, but serious adverse effects have been uncommon. Terbinafine decreases cyclosporine levels. Cimetidine increases and rifampin decreases terbinafine levels in blood.

IMIDAZOLES AND TRIAZOLES

General Features

The azole antifungals include imidazoles and triazoles. Fluconazole, itraconazole, and investigational azoles are all triazoles, so named because they have three nitrogens in the ring structure. This class has less impact on human hormonal synthesis and less hepatotoxicity than the only widely used systemic imidazole, ketoconazole. Itraconazole has many structural features in common with ketoconazole; however, it has a broader spectrum of activity and has largely replaced ketoconazole.

Azole interactions with any one class of drugs, such as benzodiazepines, HMG-CoA reductase inhibitors, or drugs that decrease gastric acidity should be considered to apply to all drugs of that class until proven otherwise. Fluconazole differs substantially from itraconazole: unlike that of itraconazole, the absorption of fluconazole is independent of food or gastric acid, and fluconazole has much less effect on the hepatic metabolism of other drugs than does itraconazole. High fluconazole blood levels engendered by azotemia or by dosages above those used in pharmacologic studies may lead to new and profound drug interactions.

All azoles have the potential for embryotoxicity and teratogenicity. In fact, it seems likely that azoles should not be given during pregnancy without a discussion of the serious risks and possible benefits with the mother. Four infants born to mothers taking at least 400 mg of fluconazole daily for coccidioidal meningitis have had severe bone, craniofacial, or cardiac abnormalities. Similarity of these abnormalities to those in pregnant animals given fluconazole suggests that fluconazole caused the defects.

Itraconazole

Itraconazole is useful in the treatment of blastomycosis, histoplasmosis, candidiasis, coccidioidomycosis, sporotrichosis, pseudallescheriasis, onychomycosis, ringworm, tinea versicolour, and some cases of aspergillosis. Its efficacy in mycoses of the central nervous system has been modest at best. Almost no bioactive drug appears in urine. Itraconazole is metabolized in the liver, with the hydroxy metabolite accounting for at least half of the antifungal activity in serum. Food increases absorption of itraconazole capsules by about threefold but substantially reduces absorption of the cyclodextrin suspension. Ability of the suspension to exert a topical as well as a systemic effect probably accounts for its improved efficacy in oropharyngeal candidiasis. The usual dosage of either oral itraconazole formulation is 100 to 200 mg once daily for oropharyngeal and esophageal candidiasis. For deep infections, itraconazole capsules are given at an initial dosage of 600 to 800 mg daily for 3 days and a subsequent dosage of 200 to 400 mg once daily continued for 6 to 12 months. Itraconazole blood levels are helpful in documenting absorption of oral itraconazole when the drug is used for the treatment of deep mycoses. An intravenous formulation is commercially available and should be considered for initial therapy in hospitalized patients in whom itraconazole absorption may be suboptimal. Immunosuppressed patients with rapidly progressing pseudallescheriasis, an infection that does not respond to amphotericin B, are candidates for intravenous itraconazole treatment.

Fluconazole

This triazole can be administered in tablet form, as a suspension, or as an intravenous infusion. With a half-life of about 31 h, fluconazole can be given once a day.

Approximately 80 per cent of the drug is excreted unchanged in the urine. Patients with creatinine clearance rates of 21 to 50 mL/min and 11 to 20 mL/min should have their fluconazole doses reduced by 50 and 75 per cent, respectively. The drug penetrates the CSF and other body fluids very well.

Nausea and abdominal distress are the most common forms of dose-limiting fluconazole toxicity. An allergic rash may develop and is particularly common among patients infected with HIV. Fatal cases of Stevens-Johnson syndrome have been described in the HIV-infected population. Alopecia commonly follows prolonged administration of ^{3}400 mg daily but resolves when therapy is discontinued. Rare cases of anaphylaxis, hepatic necrosis, and neutropenia have been described.

Fluconazole is useful in the treatment of oropharyngeal and esophageal candidiasis in adults. A single 150-mg tablet is effective in vulvovaginal candidiasis. Catheter-acquired candidemia in the immunocompetent host responds to 400 mg of fluconazole daily in conjunction with the removal of the infected catheter. Treatment should be continued for 10 to 14 days after the patient has become afebrile. Fluconazole is also effective in initial and maintenance therapy for cryptococcal meningitis in patients with AIDS, although most of these patients should initially receive a 2-week course of intravenous amphotericin B. Patients with coccidioidal meningitis can often be given fluconazole rather than intrathecal amphotericin B as maintenance therapy.

The incidence of deep candidiasis among recipients of allogeneic bone marrow transplants can be reduced by the administration of fluconazole (400 mg daily) for 75 days after initiation of the transplantation-preparative regimen. Prophylaxis in other neutropenic patients has not appeared useful. Fluconazole (200 mg daily) reduced the incidence of cryptococcosis and mucosal candidiasis among AIDS patients whose CD4+ cell counts were ?200/uL and was particularly effective among those with counts of ?50/uL. However, this regimen is not recommended because it does not reduce mortality, is expensive, and can lead to drug resistance.

Fluconazole is less effective than itraconazole in blastomycosis, histoplasmosis, and sporotrichosis. The drug is not active in aspergillosis or mucormycosis.

Amphotericin B A colloidal deoxycholate complex of the polyene drug amphotericin B is available for intravenous or intrathecal administration. In-line filters with a 0.22-um pore diameter may trap some of the colloid. The catabolism of amphotericin B is extremely slow and is not influenced by renal failure, hepatic failure, or hemodialysis. The drug's penetration into CSF and vitreous humour is poor; however, the concentrations in pleural, peritoneal, and articular exudates are adequate for many mycoses. Histoplasmosis, blastomycosis, paracoccidioidomycosis, candidiasis, and cryptococcosis are the most responsive mycoses; coccidioidomycosis, extraarticular sporotrichosis, aspergillosis, and mucormycosis are less responsive; and chromoblastomycosis, mycetoma, and pseudallescheriasis respond little, if at all. The usual course is 0.5 to 0.7 mg/kg daily for 8 to 10 weeks. Infusions are generally given in 5 per cent dextrose over 2 to 4 h.

Initial doses of amphotericin B occasionally cause marked febrile reactions that may be poorly tolerated by adult patients with limited cardiac or pulmonary function. It may be prudent to give such patients an initial 1-mg test dose followed by rapidly escalating doses, depending on tolerance. Premedication with aspirin or acetaminophen or the addition of hydrocortisone (25 mg) to the infusion decreases chills and fever. Azotemia during treatment is usual, the extent depending on the daily dose. Saline infusions have been advocated to reduce azotemia. Permanent loss of renal function is related to the total dose of amphotericin B; this condition is generally noted in adults. Other side effects include anemia, hypokalemia, renal tubular acidosis, nausea, anorexia, weight loss, phlebitis, and occasionally hypomagnesemia. Intrathecal amphotericin B has been used in coccidioidal meningitis and refractory cryptococcal meningitis, although this therapy is associated with considerable toxicity.

Three lipid formulations of amphotericin B are commercially available in the United States: amphotericin B lipid complex (ABLC), amphotericin B colloidal dispersion (ABCD), and liposomal amphotericin B (L-AB). All cause less nephrotoxicity than the older amphotericin B deoxycholate complex (ABD). Acute, febrile infusion-related reactions occur with all three lipid formulations but are most severe with ABCD. The recommended duration for initial infusions of ABCD is 1 mg/kg per hour, somewhat slower than the 2-h duration of ABLC or L-AB infusions, with the intent of decreasing febrile reactions. Premedication with acetaminophen is also an option. Use of these remarkably expensive formulations should be confined to patients who cannot tolerate the nephrotoxicity of ABD. Although the lipid formulations are also approved for patients failing to respond to ABD, there is no indication that these formulations are more effective than ABD for any mycosis.

Flucytosine Flucytosine (5-fluorocytosine) is a syntheic oral drug useful in cryptococcosis, candidiasis, and chromoblastomycosis. Within the fungal cell, flucytosine is converted to the antimetabolite 5-fluorouracil. Drug resistance appears rather rapidly when flucytosine is used alone. For this reason, the drug is generally used in combination with amphotericin B. The usual dose of flucytosine is 25 to 37.5 mg/kg every 6 h. Flucytosine is well absorbed from the gastrointestinal tract. The drug penetrates well into the CSF and is excreted unchanged in the urine. Even modest reductions in renal function may elevate flucytosine blood levels into the toxic range. Elevated levels are associated with a significant incidence of neutropenia and thrombocytopenia and also seem to predispose to colitis, the other major toxic effect of this drug. Hepatotoxicity is idiosyncratic and uncommon. An allergic rash may develop.

COCCIDIOIDOMYCOSIS

Etiologic Agent

Coccidioides immitis has two forms, growing as a white fluffy mold on most culture media but as a nonbudding spherical form (a spherule) in host tissue or under special conditions. The organism reproduces in host tissue by forming small endospores within mature spherules. After rupture of the spherule, the released endospores enlarge, become spherules, and repeat the cycle. The fungus is identified by its appearance and by the formation of thick-walled, barrel-shaped spores, called *arthrospores*, in the hyphae of the mold form.

Epidemiology, Pathogenesis and Pathology

C. immitis is a soil saprophyte found in certain arid regions of the United States, Mexico, Central America, and South America. Within the United States, most cases of infection with *C. immitis* are acquired in California, Arizona, and western Texas. A few cases are acquired by exposure to fomites from endemic areas.

Infection in humans and animals results from inhalation of wind-borne arthrospores from soil sites. This primary pulmonary infection is symptomatic in only 40 per cent of cases, with symptoms ranging from a mild influenza-like illness to severe pneumonia. Mild self-limited infections may come to medical attention because of case clusters or hypersensitivity reactions: erythema nodosum, erythema multiforme, toxic erythema, arthralgia, arthritis, conjunctivitis, or episcleritis. Case clusters occur 10 to 14 days after a group of susceptible individuals is exposed to dust in an endemic area through such activities as archaeologic excavation, rock hunting, military maneuvers, or construction work. Windstorms can carry spores to adjacent nonendemic areas and cause case clusters. The usual course of primary pneumonia is complete healing, although an area of pneumonitis (detected on radiographs) may heal by the formation of a coinlike lesion called a *coccidioidoma*. Less commonly, a single thin-walled cavity remains as a chronic sequela in the area of consolidation. Alternatively, an area of consolidation may persist as chronic pneumonia or progress to fibronodular cavitary disease. Pleural effusion may be the only manifestation of primary infection. Spontaneous healing of this form is common.

An uncommon but dreaded complication of coccidioidomycosis is dissemination beyond the lung and hilar lymph nodes. Dissemination is especially frequent among blacks, Filipinos, Native Americans, Mexican-Americans, pregnant women, and immunosuppressed patients, including those with AIDS. *C. immitis* incites a chronic granulomatous reaction in host tissue, often with caseation necrosis. Lung and hilar node lesions may show calcification. Both IgM and IgG antibodies to *C. immitis* are induced by infection, but neither type of antibody appears to be protective. The amount of specific IgG antibody is a rough measure of the antigenic mass (i.e., of the intensity of infection), and a high titer is a poor prognostic sign. Appearance of delayed hypersensitivity to antigens of *C. immitis* is most common in clinical forms of disease with a good prognosis, such as self-limited primary pulmonary disease. In skin tests for *Coccidioides* antigens, about half of patients with disseminated disease have negative results that portend a poor outcome.

Clinical Manifestations

Symptomatic primary pulmonary infection is manifested by fever, cough, chest pain, malaise, and sometimes the hypersensitivity reactions listed above. Chest radiographs may show an infiltrate, hilar adenopathy, or pleural effusion. Mild peripheral-blood eosinophilia may be found. Spontaneous improvement begins after several days to 2 weeks of illness and usually culminates in complete recovery.

The symptoms of a chronic thin-walled cavity include cough or hemoptysis in half of cases; the other half are asymptomatic. Chronic progressive pulmonary coccidioidomycosis causes cough, sputum production, variable degrees of fever, and weight loss. The first indications of dissemination usually appear during primary infection. Reactivation with dissemination in later years occurs occasionally, especially if Hodgkin's disease, non-Hodgkin's lymphoma, renal transplantation, AIDS, or immunosuppression of some other etiology has supervened. Dissemination should be suspected when fever, malaise, hilar or paratracheal lymphadenopathy, elevated sedimentation rate, and high complement fixation titèrs signal abnormal persistence in patients with primary pulmonary coccidioidomycosis. With time, lesions appear in the bone, skin, subcutaneous tissue, meninges, joints, and other sites. Chronic meningitis presents as headache of indolent onset, with or without other signs of disseminated coccidioidomycosis. Cultures and smears of cerebrospinal fluid (CSF) are most often negative, but antibody is usually detectable in CSF by complement fixation. Skin lesions are indolent and maculopapular; soft tissue and bony lesions contain pus and may present as a draining sinus. Without treatment, disseminated coccidioidomycosis progresses to death over weeks to years.

Disseminated coccidioidomycosis can progress rapidly in patients with advanced HIV infection. Fever with skin or bone lesions may be the first sign. Those who present with diffuse pulmonary infiltrates have a poor prognosis. Blood cultures are positive late in the disease, if at all.

Diagnosis

When coccidioidomycosis is suspected, sputum, urine, and pus should be examined for C. immitis by wet smear and culture. The laboratory request should indicate clearly that coccidioidomycosis is suspected, because the mold form must be handled with extreme care to prevent infection of laboratory personnel. On biopsy, smaller spherules must be distinguished from nonbudding forms of Blastomyces and Cryptococcus, but the appearance of the mature spherule is diagnostic.

Serologic tests are very helpful in the diagnosis of coccidioidomycosis. Latex agglutination and agar gel diffusion tests are useful in screening sera for antibody to *Coccidioides*. The complement fixation test is used for CSF determinations and for the confirmation and quantitation of serum antibody detected by screening tests. The number of cases with a positive complement fixation test depends on the severity of disease and on the laboratory performing the test. Positive tests are least common among patients with solitary pulmonary cavities or primary pulmonary infection, while sera from patients with disseminated disease in multiple organs are nearly all positive. Seroconversion is helpful in primary pulmonary coccidioidomycosis but may not occur for up to 8 weeks after onset. A positive complement fixation test of unconcentrated CSF is diagnostic of meningitis. Rarely, a parameningeal focus causes a positive complement fixation test of CSF.

Conversion of the skin test from negative to positive with spherulin may take place between days 3 and 21 of symptoms in primary pulmonary coccidioidomycosis. Skin testing can be helpful in epidemiologic studies, such as investigations of case clusters or the definition of endemic areas. The utility of skin testing as a diagnostic tool is limited by the persistence of positive tests resulting from remote exposures to *Coccidioides* and by the frequency of negative skin tests among patients with either thin-walled cavities or disseminated coccidioidomycosis. A positive skin test has not predicted dissemination in HIV-infected patients. The presence of complement-fixing antibody to *C. immitis* in AIDS patients should prompt a search for active infection.

Treatment

Primary pulmonary coccidioidomycosis usually resolves spontaneously. Some physicians give a few weeks of treatment with intravenous amphotericin B or itraconazole to patients with unusually severe or protracted primary infection in the hope of aborting disseminated or chronic pulmonary disease.

Patients with severe or rapidly progressing disseminated coccidioidomycosis are first given intravenous amphotericin B at a dose of 0.5 to 0.7 mg/kg daily. Patients whose condition improves after 2 to 3 months of treatment with amphotericin B or who have more indolent disseminated infection are given itraconazole (200 mg twice daily) or fluconazole (400 to 600 mg/d). These oral agents are useful for long-term suppression of infection, and treatment should be continued for years. Patients with coccidioidal meningitis usually are initially given fluconazole (400 to 800 mg/d) but may require intrathecal amphotericin B. Hydrocephalus is a frequent complication of uncontrolled meningitis. Surgical debridement of bone lesions or drainage of abscesses can be helpful. The prognosis for ultimate cure of disseminated coccidioidomycosis is guarded.

Resection of chronic progressive pulmonary lesions is a helpful adjunct to chemotherapy when infection is confined to the lung and to one lobe. A single thin-walled cavity tends to close spontaneously and ordinarily is not resected. Such a cavity responds poorly to chemotherapy.

BLASTOMYCOSIS

Etiologic Agent

Blastomyces dermatitidis is a dimorphic fungus that grows at room temperature as a white or tan mold but grows within the host as budding, round yeastlike cells. The fungus can be identified on the basis of its appearance, its dimorphism, the small spores borne on hyphae of the mold form, or the results of nucleic acid hybridization. When isolates of the two opposite mating types are grown close together on special culture medium, such as yeast extract or soil extract agar, sporulating structures that characterize the perfect state (teleomorph), called *Ajellomyces dermatitidis*, appear.

Epidemiology

The infection is restricted by geography and age. Blastomycosis is uncommon in any locality, but most cases occur in the southeastern, central, and mid-Atlantic areas of the United States, with occasional cases in other localities in the United States and Canada. Cases have also been encountered in Africa, Mexico, Central America, and (rarely) South America. Most patients are between 20 and 69 years old. The male-to-female ratio is about 10:1. There is no occupational predisposition to the development of blastomycosis.

Pathogenesis and Pathology

Infection with *B. dermatitidis* appears to be acquired by inhalation of the fungus from soil, decomposed vegetation, or rotting wood. Several case clusters have resulted from participation in recreational activities in wooded areas along waterways. Infection is not transmissible from person to person. The initial pulmonary infection may either heal spontaneously or become chronic. Spread to other portions of the lung, cavitation, or endobronchial lesions may be found in patients with chronic disease. Whether or not the lung lesion resolves spontaneously, infection commonly spreads hematogenously to the skin, subcutaneous tissue, bone, prostate, epididymis, or mucosa of the nose, mouth, or larynx. Less commonly, infection spreads to the brain, meninges, liver, lymph nodes, or spleen. Dissemination may not be evident for weeks or years after the appearance of the lung lesion. Progressive infection is only rarely attributable to an underlying disease, to HIV infection, or to immunosuppressive treatment. The inflammatory response includes lymphocytes, giant cells, and neutrophils. Pseudoepitheliomatous hyperplasia may be striking and may lead to a mistaken diagnosis of squamous cell carcinoma.

Clinical Manifestations

A few patients have acute, self-limited pneumonia. Fever, productive cough, myalgia, and malaise usually resolve within a month. Pulmonary infiltrates clear slowly as *B. dermatitidis* disappears from the sputum.

In the vast majority of patients, blastomycosis has an indolent onset and a chronically progressive course. Fever, cough, weight loss, lassitude, skin lesions, and chest ache are common. Skin lesions favour exposed areas and enlarge over many weeks from pimples to well-circumscribed, verrucous, crusted, or ulcerated lesions. Pain and regional lymphadenopathy are minimal. Large chronic lesions may undergo central healing with scarring and contracture. Mucous membrane lesions resemble squamous cell carcinoma. Chest x-ray findings are abnormal in two-thirds of patients, with one or more pneumonic or nodular infiltrates. Calcification, hilar adenopathy, and large pleural effusions are rare. Osteolytic lesions may be found in nearly any bone and present as a cold abscess or a draining sinus. Extension to a contiguous joint may cause indolent swelling, pain, and restricted motion. Prostatic and epididymal lesions clinically resemble those of tuberculosis.

Diagnosis

The diagnosis of blastomycosis is made by demonstration of the fungus in a culture of sputum, pus, or urine. An expert can diagnose blastomycosis on the basis of the appearance of the organism in wet smear or histopathologic section. The fungus may be visible in a sputum cytology smear but is easily overlooked.

Treatment

A few patients have developed only transitory lung lesions, but no guidelines are known to distinguish these patients from those whose disease will progress locally or disseminate. Therefore, every patient should receive treatment. Intravenous amphotericin B is the drug of choice for patients with rapidly progressive infections, severe illness, or central nervous system lesions. Skin and noncavitary lung lesions should be treated for about 8 to 10 weeks. The recommended total dose for an adult is about 2 g. Cavitary lung disease or infection extending beyond the lung and skin should be treated for about 10 to 12 weeks.

Oral itraconazole (200 mg twice daily with food) is the drug of choice for the treatment of patients who have indolent nonmeningeal blastomycosis of mild to moderate severity and who take the drug reliably. Therapy with itraconazole is continued for 6 to 12 months.

CRYPTOCOCCOSIS

Etiologic Agent

Cryptococcosis is an infection caused by the yeastlike fungus *Cryptococcus neoformans*. This fungus reproduces by budding and forms round, yeastlike cells. Within the host and on certain culture media, a large polysaccharide capsule surrounds each yeast cell. The fungus grows well in smooth, creamy-white colonies on Sabouraud's or other simple media. Identification of the organism is based on gross and microscopic appearance, biochemical test results. The results of nucleic acid hybridization or the formation of brown pigment on Niger seed agar can also be used for identification.

The fungus has four capsular serotypes, designated A, B, C, and D. There are also two mating types. Coculture of opposite mating types creates a transient diploid state called *Filobasidiella neoformans* var. neoformans for serotypes A and D and *F. neoformans* var. bacillispora for serotypes B and C. Organisms not cultured under mating conditions are designated *C. neoformans* var. neoformans for serotypes A and D and *C. neoformans* var. gattii for serotypes B and C; a simple colour medium distinguishes the two varieties.

Epidemiology

Weathered pigeon droppings commonly contain serotype A or D (*C. neoformans* var. *neoformans*). *C. neoformans* var. *gattii* has been isolated from the litter around trees of the species *Eucalyptus camaldulensis* and *E. tereticornis*. Eucalyptus isolates have so far typed as serotype B. The distribution of these eucalyptus species in Australia corresponds to the distribution of infections due to *C. neoformans* var. *gattii* in that country. The high prevalence of these trees in other subtropical climates has been postulated to explain the relative restriction of such infections to warm climates.

Cryptococcosis due to *C. neoformans* var. *neoformans* is a common complication of late infection with HIV. The incidence appears to be declining in some areas of the United States, probably as a result of highly active antiretroviral therapy and use of fluconazole for oropharyngeal candidiasis. Patients who have undergone solid-organ transplantation or glucocorticoid therapy and those with sarcoidosis are also at increased risk for infections with *C. neoformans* var. *neoformans*. Almost all such infections are caused by serotype A, although serotype D occurs in up to 20 per cent of cases in Western Europe. Infections with var. *gattii* have been rare among AIDS patients and other immunocompromised patients, even in subtropical climates, where var. *gattii* infection occurs in previously healthy individuals.

Animals, particularly cats, can acquire cryptococcosis but have not transmitted the infection to other animals or to humans. The source from which humans acquire the infection is unknown, with the rare exception of cases acquired through a transplanted cornea, kidney, or other solid organ. Cryptococcosis is rare before puberty.

Pathogenesis and Pathology

Infection is thought to be acquired by inhalation of fungus into the lungs. Pulmonary infection has a tendency toward spontaneous resolution and is frequently asymptomatic. Silent hematogenous spread to the brain leads to clusters of cryptococci in the perivascular areas of cortical gray matter, in the basal ganglia, and, to a lesser extent, in other areas of the central nervous system. The inflammatory response around these foci is usually scant. In the more chronic cases, a dense basilar arachnoiditis is typical. Lung lesions are characterized by intense granulomatous inflammation. Cryptococci are best seen in tissue by staining with methenamine silver or periodic acid-Schiff. Although a strongly positive result on mucicarmine staining of tissue is diagnostic, staining varies from intense to absent.

Clinical Manifestations

Most patients have *meningoencephalitis* at the time of diagnosis. This form of the infection is invariably fatal without appropriate therapy; death occurs any time from 2 weeks to several years after the onset of symptoms. Early manifestations include headache, nausea, staggering gait, dementia, irritability, confusion, and blurred vision. Both fever and nuchal rigidity are often mild or lacking. Papilledema is evident in one-third of cases at the time of diagnosis. Cranial nerve palsies, typically asymmetric, occur in about one-fourth of cases. Other lateralized signs are rare. With progression of the infection, deepening coma and signs of brainstem compression appear. Autopsy often reveals cerebral edema in more acute cases and hydrocephalus in more chronic cases.

Pulmonary cryptococcosis causes chest pain in about 40 per cent of patients and cough in 20 per cent. The chest x-ray shows one or denser infiltrates, which are often well circumscribed. Cavitation, pleural effusions, and hilar adenopathy are infrequent. Calcification is not evident, and fibrotic stranding is rarely noticeable.

Ten percent of patients with cryptococcosis have skin lesions, and the vast majority of patients with skin lesions have disseminated infection. One or a few asymptomatic tiny papular lesions appear and slowly enlarge; they display a tendency toward central softening leading to ulceration. Osteolytic lesions occur in 4 per cent of cases and usually present as a cold abscess. Rare manifestations of cryptococcosis include prostatitis, endophthalmitis, hepatitis, pericarditis, endocarditis, and renal abscess.

Diagnosis

Fever and headache in a patient with AIDS or with risk factors for HIV infection suggest the possibility of cryptococcosis, toxoplasmosis, or central nervous system lymphoma. Evidence of a focal lesion on magnetic resonance imaging is unusual in cryptococcosis. Most cryptococcal cerebral mass lesions occur in patients infected with C. neoformans var. gattii who also have meningitis. In patients without AIDS, meningitis due to C. neoformans resembles that due to Mycobacterium tuberculosis, Histoplasma capsulatum, Coccidioides immitis, or metastatic cancer. Lumbar puncture is the single most useful diagnostic test. An india ink smear of centrifuged cerebrospinal fluid (CSF) sediment reveals encapsulated yeast in more than half of cases, although artifacts can cause confusion. In patients without AIDS, levels of glucose in CSF are reduced in half of all cases; protein levels are usually increased; and lymphocytic pleocytosis is usually found. CSF abnormalities are less pronounced in patients with AIDS, although india ink smear is more often positive.

Approximately 90 per cent of patients with cryptococcal meningoencephalitis, including all those with a positive CSF smear, have capsular antigen detectable in CSF or serum by latex agglutination. An enzyme immunoassay for cryptococcal antigen is also available. Occasional false-positive results in the above tests make culture the definitive diagnostic test and have prevented serum antigen from being a useful screening test in asymptomatic patients with AIDS. *C. neoformans* is often present in urine from patients with meningoencephalitis. Fungemia occurs in 10 to 30 per cent of patients and is particularly common among patients with AIDS.

Pulmonary cryptococcosis mimics malignancy with regard to radiographic findings and symptoms. Sputum culture is positive in only 10 per cent of cases, and serum antigen tests are positive in only one-third. Occasionally, *C. neoformans* appears in one or more sputum specimens as an endobronchial saprophyte. Biopsy is usually required for diagnosis. Cutaneous cryptococcosis may be mistaken for a comedo, basal cell carcinoma, or sarcoidosis. In patients with AIDS, skin lesions may be numerous and are sometimes mistaken for molluscum contagiosum. Biopsy reveals myriad cryptococci. Osseous cryptococcosis resembles tuberculosis.

Treatment

Patients with AIDS and cryptococcosis are treated initially with intravenous amphotericin B (0.7 mg/kg daily) for at least 2 weeks and until their clinical condition is stable; thereafter, they receive fluconazole. The addition of flucytosine (25 mg/kg every 6 h) to amphotericin B for 2 weeks has minimal impact on morbidity and mortality. After treatment with amphotericin B, fluconazole (400 mg) is given once daily. Daily doses of 800 mg have been used with marginal changes in toxicity or efficacy. The addition of flucytosine to fluconazole increases gastrointestinal intolerance. After infection is controlled, treatment with a smaller dose of fluconazole (200 mg/d) is continued indefinitely. Itraconazole is less effective than fluconazole for maintenance therapy. It is not yet known whether patients whose CD4+ T lymphocyte counts have exhibited a sustained rise in response to antiretroviral therapy can safely discontinue fluconazole maintenance therapy.

In patients without AIDS, the therapeutic goal is to cure the infection, not merely to control its symptoms. A single intensive course is given until cultures from all previously positive sites (particularly CSF) become convincingly negative. Normalization of the glucose level in lumbar CSF is desirable, but complete clearing of CSF or serum antigen during therapy is not essential. Amphotericin B is the best-studied regimen. Flucytosine has been added to amphotericin B to accelerate the culture response, but grave toxicity can result unless flucytosine blood levels are kept below 100 ug/mL. Case reports have described patients without HIV infection who have responded to fluconazole or liposomal amphotericin B, but the dose and duration of treatment required to cure cryptococcal meningitis are undefined. Amphotericin B lipid complex and amphotericin B colloidal dispersion are not recommended pending further study.

Hydrocephalus may be the presenting manifestation or a later complication of cryptococcosis. Blindness, dementia, and personality change are among the other sequelae. Daily lumbar puncture or CSF shunting has been advocated in the hope of averting permanent blindness for patients with marked cerebral edema who have incipient blurred vision.

Patients with extraneural cryptococcosis most often require treatment with intravenous amphotericin B, with or without flucytosine. Observation or excision of lesions may suffice for some patients who have previously been healthy; who have a single focus in lung, skin, or bone; and who have no cryptococci in CSF, urine, or blood.

CANDIDIASIS

Etiologic Agents

Candida albicans is the most common cause of mucosal candidiasis and is responsible for about half of all cases of candidemia in hospitalized patients. A small proportion of *C. albicans* isolates have been transferred to a new species, *C.*

dubliniensis. *C. tropicalis*, *C. parapsilosis*, *C. guilliermondii*, *C. glabrata* (formerly *Torulopsis glabrata*), *C. krusei*, and a few other *Candida* species also cause potentially fatal bloodstream infection. Many of these non-*albicans* species can enter the bloodstream through an intravascular catheter. *Candida* species, taken together, are the fifth most common cause of nosocomial bloodstream infections.

All *Candida* species pathogenic for humans are also encountered as commensals of humans, particularly in the mouth, stool, and vagina. These species grow rapidly at 25° to 37°C on simple media as oval, budding cells. In special culture media and in tissue, hyphae or elongated branching structures called *pseudohyphae* are formed. *C. glabrata* differs from other members of the genus in that it forms no true hyphae or pseudohyphae in vitro or in infected tissue. *C. albicans* and *C. dubliniensis* can be identified presumptively by their ability to form germ tubes in serum or by the formation of thick-walled large spores called *chlamydospores*. Final identification of all *Candida* species requires biochemical tests.

Pathogenesis

Candidiasis is often preceded by increased colonization of the mouth, vagina, and stool with *Candida* due to broad-spectrum antibiotic therapy. Additional local and systemic factors favour infection. Oropharyngeal thrush is particularly likely to occur in neonates and in patients with diabetes mellitus, HIV infection, or dentures. Vulvovaginal candidiasis is especially common in the third trimester of pregnancy. *Candida* from the perineum can enter the urinary tract via an indwelling bladder catheter. Cutaneous candidiasis most often involves macerated skin, such as that in the diapered area of infants, under pendulous breasts, or on hands constantly in water or covered by occlusive gloves. *Candida* can pass from the colonized surface into deep tissue when the integrity of the mucosa or skin is violated by perforation of the gastrointestinal tract through trauma, surgery, or peptic ulceration or by mucosal damage due to cytotoxic agents used for cancer chemotherapy. Although *Candida* is not normally a resident of the skin, secretions from the mouth, rectum, or vagina as well as drainage from surgical wounds or tracheostomy sites can contaminate the hub or skin site of a catheter in an umbilical or central vein. Intravenous drug abuse or third-degree burns can also provide a skin portal for *Candida* that can lead to deep candidiasis. Once *Candida* has passed the integumentary barrier, very low birth weight (in neonates) and neutropenia or glucocorticoid therapy (in any patient) markedly compromise host defence. Hematogenous seeding is particularly evident in the retina, kidney, spleen, and liver.

Clinical Manifestations

Oral thrush presents as discrete and confluent adherent white plaques on the oral and pharyngeal mucosa, particularly in the mouth and on the tongue. These lesions are usually painless, but fissuring at the corners of the mouth can be painful. Unexplained oropharyngeal thrush raises the possibility of HIV infection. Oral thrush is common in acute HIV infection and becomes increasingly common as the CD4+ cell count falls. At CD4+ counts < 50/uL, esophageal thrush also becomes common. HIV infection appears not to be an independent risk factor for vulvovaginal thrush.

Cutaneous candidiasis presents as red macerated intertriginous areas, paronychia, balanitis, or pruritus ani. Candidiasis of the perineal and scrotal skin may be accompanied by discrete pustular lesions on the inner aspects of the thighs. *Chronic mucocutaneous candidiasis* or *candidal granuloma* typically presents as circumscribed hyperkeratotic skin lesions, crumbling dystrophic nails, partial alopecia in areas of scalp lesions, and both oral and vaginal thrush. Systemic infection is very rare, but disfigurement of the face and hands can be severe. Other findings may include chronic epidermophytosis, dental dysplasia, and hypofunction of the parathyroid, adrenal, or thyroid gland. A variety of defects in T cell function have been described in these patients. Vulvovaginal thrush causes pruritus, discharge, and sometimes pain on intercourse or urination. Speculum examination reveals an inflamed mucosa and a thin exudate, often with white curds.

Esophageal candidiasis is often asymptomatic but can cause substernal pain or a sense of obstruction on swallowing. Most lesions are in the distal third of the esophagus and appear on endoscopy as areas of redness and edema, focal white patches, or ulcers. Biopsy or brushing is required for diagnosis and for detection of concomitant infections, particularly herpes simplex in patients with hematologic malignancies and cytomegalovirus infection in AIDS patients. Esophagography (barium swallow) is diagnostically insensitive but may reveal spasm or mucosal irregularities. *Candida* esophagitis can cause bleeding and impaired alimentation. Hematogenous dissemination from the esophagus probably occurs in some neutropenic patients but is rarely reported in HIV-infected patients.

Candida can cause cystitis, pyelitis, or renal papillary necrosis in an obstructed urinary tract. When a colonized urinary tract is operated on or instrumented, candidemia may result. However, most patients with *Candida* cultured from the urine simply have bladder colonization from a Foley catheter or a sizable volume of residual urine. Contamination of a voided midstream specimen by vaginal *Candida* is also common.

Candidemia originating from an intravascular catheter may clear in the immunocompetent patient when the catheter is removed. Focal seeding of the retina can take place even if candidemia clears and the patient becomes afebrile. Unilateral or bilateral small white retinal exudates appear within 2 weeks of the onset of candidemia. Lesions may regress spontaneously or enlarge slowly. The vitreous humour becomes cloudy, and the patient notices blurring, ocular pain, or a scotoma. Retinal detachment, vitreous abscess, and extension to the anterior chamber can occur over the ensuing weeks. These retinal lesions, present in ~10 per cent of nonneutropenic patients with candidemia, are the principal reason that systemic antifungal therapy is recommended for all patients with candidemia. Funduscopy should be performed to be certain that retinal lesions, if present, resolve completely. Most cases with ocular involvement have occurred in nonneutropenic patients. In contrast, so-called hepatosplenic candidiasis is usually recognized in patients with acute leukemia who are recovering from profound neutropenia. This entity, better called *chronic disseminated candidiasis*, originates from intestinal seeding of the portal and venous circulation.

Fever, modestly elevated serum concentrations of alkaline phosphatase, and multiple small abscesses evident on ultrasonography, magnetic resonance imaging, or computed tomography of the liver, spleen, or kidney suggest the diagnosis. During acute candidemia in neutropenic patients, small erythematous papules may appear anywhere on the skin.If the patient does not expire promptly from disseminated candidiasis, the lesions will develop a necrotic centre. Painful muscle lesions may also be found. Punch biopsy of a skin lesion helps distinguish this extremely grave condition from *Malassezia* folliculitis, a similar-appearing but benign condition that can involve the cape area of the chest or the extremities of a sweaty febrile patient.

Hematogenous seeding in the neutropenic patient is occasionally visible radiologically as tiny pulmonary nodules. *Candida* pneumonia, apart from hematogenous candidiasis, is very rare. Organisms seeding a native or prosthetic cardiac valve originate principally from central venous catheters; occasionally, valvular seeding is encountered in intravenous drug abusers. Emboli to large arteries, such as the iliac or femoral artery, are characteristic. Intravenous injection of impure brown heroin has caused a clinical syndrome consisting of *Candida* endophthalmitis and purulent folliculitis, sometimes accompanied by vertebral osteomyelitis. This diffuse folliculitis favours hairy areas, including the scalp and bearded facial skin. *Candida* can cause indolent arthritis, most commonly of the knee, in patients who have received glucocorticoid injections into the joint, in patients who are immunosuppressed, and in low-birth-weight neonates. Prosthetic joints may become infected during implantation. Scanty growth of *Candida* from joint fluid can cause the laboratory to incorrectly dismiss the organism as a contaminant.

Hematogenous dissemination can lead to brain abscess or chronic meningitis. Diagnosis of infections of ventriculoperitoneal shunts is difficult because symptoms are indolent and cultures of lumbar fluid are usually sterile.

Diagnosis

Demonstration of pseudohyphae on wet smear with confirmation by culture is the procedure of choice for diagnosing superficial candidiasis.Scrapings for the smear may be obtained from skin, nails, and oral and vaginal mucosa. Culture alone is not diagnostic; however, recovery of *Candida* species from multiple superficial sites in immunosuppressed patients may portend visceral invasion. Deeper lesions due to *Candida* may be diagnosed by histologic section of biopsy specimens or by culture of cerebrospinal fluid, blood, joint fluid, or surgical specimens. Blood cultures are useful in the diagnosis of *Candida* endocarditis and intravenous catheter-induced sepsis but are positive less often in other forms of disseminated disease. Serologic tests for antibody or antigen are not useful.

Treatment

Cutaneous candidiasis of macerated areas responds to measures that reduce moisture and chafing plus topical application of an antifungal agent in a nonocclusive base. Nystatin powder or a cream containing ciclopirox or an azole is useful. Clotrimazole, miconazole, econazole, ketoconazole, sulconazole, and oxiconazole are available as creams or lotions. *Candida* vulvovaginitis responds better to an azole than to nystatin suppositories. There is little difference in efficacy among miconazole, clotrimazole, tioconazole, butoconazole, and terconazole vaginal formulations. Systemic treatment of *Candida* vulvovaginitis with a single 150-mg capsule of fluconazole is more convenient than topical treatment but also poses a higher risk of adverse effects. Clotrimazole troches, used five times a day, are more effective in oral and esophageal candidiasis than nystatin suspension. Oral fluconazole (100 to 200 mg once daily) is more convenient and more effective in esophagitis than clotrimazole troches. Esophagitis not responding to fluconazole may warrant repeat endoscopy to exclude other conditions. Itraconazole suspension (100 to 200 mg/d) alleviates *Candida* esophagitis in some patients in whom fluconazole treatment fails. Amphotericin B suspension has limited use but can be tried in patients whose oropharyngeal candidiasis does not respond to azoles.

Management of recurrent oropharyngeal candidiasis in the HIV-infected patient presents special problems. Patients with CD4+ cell counts < 100/uL who have received prolonged fluconazole therapy are at risk of developing azole resistance, requiring an increased dose to mount a response, relapsing early, and eventually failing to respond well to any dose of fluconazole. The increasing azole resistance in this population suggests that HIV-infected patients with oropharyngeal candidiasis should be treated for each individual episode and that only when episodes become intolerably frequent should weekly or daily preventive therapy be given and even then at the lowest dose required to maintain remission. In contrast, AIDS patients with *Candida* esophagitis are so prone to relapse that preventive therapy with fluconazole is recommended for all proven cases. Most HIV-infected patients with azole-resistant oropharyngeal candidiasis also have esophagitis. Nearly all patients with azole-resistant oropharyngeal or esophageal candidiasis respond to intravenous amphotericin B (0.3 to 0.5 mg/kg daily) but relapse promptly after the completion of therapy.

Bladder thrush responds to bladder irrigations with amphotericin B (50 ug/mL for 5 days). If no bladder catheter is in place, oral fluconazole can be used to control candiduria. In all forms of superficial candidiasis, relapse after successful treatment is common unless the underlying factor can be eliminated.

Intravenous amphotericin B is the drug of choice in disseminated candidiasis. The deoxycholate formulation is usually given at a dosage of 0.5 to 0.7 mg/kg daily. Open, noncomparative studies of the lipid formulations of amphotericin B have suggested that they may be useful in disseminated candidiasis, but the optimal dose and formulation remain unknown. Fluconazole in an adult dose of 100 mg/d is probably the drug of choice for chronic mucocutaneous candidiasis.

In immunocompetent patients with intravenous catheter-acquired *C. albicans* fungemia, the catheter should be removed in conjunction with the administration of either fluconazole (400 mg/d) or amphotericin B (0.5 mg/kg daily). Patients with suppurative phlebitis of a peripheral vein should have the infected portion of the vein excised. Therapy for candidemia is continued for 2 weeks after the patient becomes afebrile. The *Candida* species involved should be considered in choosing between fluconazole and amphotericin B. *C. krusei* and *C. inconspicua* are rare causes of candidemia but are resistant to fluconazole in vitro. *C. glabrata* exhibits intermediate susceptibility to fluconazole, but too few cases have been studied to determine whether candidemia involving that species will respond as well to fluconazole as to amphotericin B. Strains of *C. lusitaniae* resistant to amphotericin B but susceptible to azoles have been encountered. Intravenous amphotericin B, with or without flucytosine, is the preferred treatment for *Candida* endophthalmitis, although cures have been reported with fluconazole. Pars plana vitrectomy may facilitate diagnosis and cure when a *Candida* vitreous abscess is present. Injection of amphotericin B into the vitreous humour can also be helpful.

Injection of amphotericin B into an infected joint, pleural cavity, or peritoneum is rarely indicated. Removal of prostheses, including prosthetic joints, cardiac valves, peritoneal dialysis catheters, and central venous catheters, is usually essential. Collections of pus, such as those in the postoperative abdomen, need to be drained surgically or by percutaneous, computed tomography-guided catheterization; an exception relates to the numerous small abscesses in liver, spleen, or kidney in chronic disseminated candidiasis, which cannot be drained effectively and require prolonged antifungal therapy. In general, treatment should continue until the patient with chronic disseminated candidiasis has been afebrile and nonneutropenic for at least 2 weeks. Defects may persist on imaging studies long after cure. Relapse during another episode of neutropenia is common unless the patient is receiving amphotericin B. Repeat cytotoxic therapy or even bone marrow transplantation can be undertaken in patients with prior chronic disseminated candidiasis, but amphotericin B should be given prophylactically during neutropenia.

Fluconazole can decrease the incidence of deep candidiasis in recipients of allogeneic bone marrow transplants when 400 mg is given daily until engraftment. Although the incidence of superficial candidiasis is also decreased by fluconazole prophylaxis, superficial infection can be readily detected and treated. Aspergillosis is not prevented by prophylactic fluconazole. Studies of leukemic and other neutropenic patients have found no beneficial effect of prophylactic fluconazole.

ASPERGILLOSIS

Etiologic Agents

Aspergillus fumigatus is the most common cause of aspergillosis, but *A. flavus, A. niger*, and several other species can also cause disease. *Aspergillus* is a mold with septate hyphae about 2 to 4 um in diameter. The fungus is identified by its gross and microscopic appearance in culture.

Pathogenesis and Pathology

All the common species of *Aspergillus* that cause disease in humans are ubiquitous in the environment, growing on dead leaves, stored grain, compost piles, hay, and other decaying vegetation. Inhalation of *Aspergillus* spores must be extremely common, but disease is rare. Invasion of lung tissue is confined almost entirely to immunosuppressed patients, in roughly 90 per cent of whom two of the following three conditions will be operative: a granulocyte count in peripheral blood of < 500/uL, treatment with supraphysiologic doses of adrenal glucocorticoids, and a history of treatment with cytotoxic drugs such as cyclosporine. Invasive aspergillosis is an occasional complication of AIDS. *Aspergillus* infection is characterized by hyphal invasion of blood vessels, thrombosis, necrosis, and hemorrhagic infarction. Chronic granulomatous disease of childhood also predisposes to invasive pulmonary aspergillosis, but in that situation the inflammatory response is a pyogranuloma and blood vessel invasion is rare.

Massive inhalation of *Aspergillus* spores by healthy persons can lead to acute, diffuse, self-limited pneumonitis. Epithelioid granulomas with giant cells and central pyogenic areas containing hyphae are detected in these cases. Spontaneous recovery taking several weeks is the usual course. *Aspergillus* can colonize the damaged bronchial tree, pulmonary cysts, or cavities of patients with underlying lung disease. Balls of hyphae within cysts or cavities (aspergillomas), usually in the upper lobe, may reach several centimeters in diameter and may be visible on chest x-ray. Tissue invasion does not occur. The term *allergic bronchopulmonary aspergillosis* denotes the condition of patients with preexisting asthma who have eosinophilia, IgE antibody to *Aspergillus*, and fleeting pulmonary infiltrates from bronchial plugging.

Clinical Manifestations

Endobronchial saprophytic pulmonary aspergillosis presents as chronic productive cough, often with hemoptysis, in a patient with prior chronic lung disease, such as tuberculosis, sarcoidosis, bronchiectasis, or histoplasmosis. *Aspergillus* may be spread from its endocavitary or endobronchial site to the pleura during the course of bacterial lung abscess or surgery. Patients reported to have chronic necrotizing *Aspergillus* pneumonia appear in most instances to have had saprophytic endobronchial colonization and a pulmonary process attributable to another disease, with or without superimposed bacterial infections. Patients with chronic pneumonia and *Aspergillus* in the sputum should be assumed to have either pneumonia of a different etiology (e.g., histoplasmosis) or *Aspergillus* pneumonia with underlying immunosuppression (e.g., chronic granulomatous disease or infection with HIV).

Invasive aspergillosis in the immunocompromised host presents as an acute, rapidly progressive, densely consolidated pulmonary infiltrate and is most common among patients with acute leukemia and recipients of tissue transplants. Infection progresses by direct extension across tissue planes and by hematogenous dissemination to lung, brain, and other organs. Computed tomography (CT) has been particularly valuable in suggesting the diagnosis of invasive pulmonary aspergillosis in patients with neutropenia. The earliest CT finding is one or more small pulmonary nodules. As a nodule enlarges, the dense central core of infarcted tissue becomes surrounded by edema or hemorrhage, forming a hazy rim called the *halo sign*. This rim disappears in a few days as the dense core enlarges. When bone marrow function recovers, the infarcted central core cavitates, creating the *crescent sign*. *Aspergillus* may invade immunosuppressed patients through the skin at a site of minor trauma or through the upper airway mucosa. Early lesions in the nose should be sought in patients with neutropenia who have fever and minimal epistaxis. Scarlet-red patches of the mucosa rapidly become necrotic and white, then black. Rapid extension into the adjacent paranasal sinus, orbit, or face is usual, with or without the appearance of lung lesions.

Aspergillus sinusitis in immunocompetent patients may take three forms. A ball of hyphae may form in a chronically obstructed paranasal sinus, without tissue invasion. Much less commonly, a chronic, fibrosing granulomatous inflammation associated with *Aspergillus* hyphae within tissue may begin in the sinus and spread slowly to the orbit and the brain. *Aspergillus* is also a cause of allergic fungal sinusitis, but dark-walled fungi (e.g., *Cladosporium, Alternaria*) are more common in this setting. Patients usually have a history of chronic allergic rhinitis, sometimes with nasal polyps, but are otherwise healthy, presenting with painless proptosis, nasal obstruction, or dull aching pain. On CT or magnetic resonance imaging, a solid soft tissue mass pushing out the lateral wall of the ethmoid sinus or the medial wall of the maxillary sinus may be detected. On sinus exploration, the mucosa is found to be thickened and inflamed but intact. Within the sinus cavity, sticky mucopus with strands of neutrophils, eosinophils, Charcot-Leyden crystals, and occasional hyphae can be found.

Aspergillosis in HIV-infected patients most commonly involves the lung, presenting as fever, cough, and dyspnea. Typically, the CD4 cell count is below 50/uL. Roughly half of these patients have neutropenia or have recently been treated with glucocorticoids. Bilateral diffuse or focal pulmonary infiltrates with a tendency to cavitate constitute the most common radiologic manifestation. Well-localized, white, necrotic pseudomembranes full of hyphae or ulcers may develop in the trachea or the major bronchi. Progression of bronchitis to pneumonia is usual, but hematogenous dissemination is uncommon. Either allergic or invasive *Aspergillus* sinusitis can occur in HIV-infected patients; the allergic form can develop even at CD4 cell counts above 50/uL.

The growth of *Aspergillus* on cerumen and detritus within the external auditory canal is termed *otomycosis*. Trauma to the cornea may cause *Aspergillus* keratitis. Endophthalmitis follows the introduction of *Aspergillus* into the globe by trauma or surgery. *Aspergillus* may infect intracardiac or intravascular prostheses.

Diagnosis

The repeated isolation of *Aspergillus* from sputum or the demonstration of hyphae in sputum or bronchoalveolar lavage fluid suggests endobronchial colonization or infection. Even a single isolation of *Aspergillus* from the sputum of a neutropenic patient with pneumonia, particularly a child or a nonsmoker, suggests the diagnosis of invasive aspergillosis. In patients with advanced AIDS, fever, and cough, the isolation of *Aspergillus* from respiratory secretions raises the possibility of aspergillosis and thus should prompt bronchoscopy. Fungus ball of the lung is usually detectable by chest x-ray. IgG antibody to *Aspergillus* antigens is demonstrable in the serum of many colonized patients and of virtually all patients with fungus ball.

Biopsy is usually required for the diagnosis of invasive aspergillosis of the lung, nose, paranasal sinus, bronchi, or sites of dissemination. Blood cultures are rarely positive, even in patients with infected cardiac valves (native or prosthetic). Detection of galactomannan antigen in serum suggests the diagnosis, but false-positives are frequent, particularly in children. *Aspergillus* hyphae can be identified presumptively by histology, but culture is required for confirmation and for determination of the species. Only culture can reliably distinguish aspergillosis from pseudallescheriasis; drug therapy for these two diseases differs.

Treatment

Patients with severe hemoptysis due to fungus ball of the lung may benefit from lobectomy. Poor pulmonary function in residual lung and dense pleural adhesions around the lesion can complicate the resection. Systemic chemotherapy is of no value in endobronchial or endocavitary aspergillosis.

Treatment with intravenous amphotericin B (1.0 to 1.5 mg/kg daily) has resulted in the arrest or cure of invasive aspergillosis when immunosuppression is not severe. Liposomal amphotericin B at daily doses of 1 to 4 mg/kg has given results that seem roughly comparable to those obtained with amphotericin B deoxycholate. Itraconazole (200 mg twice daily) is useful in some less immunosuppressed patients with indolent or slowly progressive invasive aspergillosis. Surgery is the only treatment needed for fungus ball of the sinus and for allergic fungal sinusitis. Antifungal therapy has little effect on either entity if used alone, but chronic suppressive therapy has been begun postoperatively for relapse of allergic fungal sinusitis. The prognosis for cure of invasive aspergillosis in the paranasal sinus is very poor when the patient has profound and unremitting neutropenia. The prognosis is better in less immunosuppressed patients.

MUCORMYCOSIS

Etiologic Agents

Species of *Rhizopus*, *Rhizomucor*, and *Cunninghamella* are the most common causes of mucormycosis, but species of *Apophysomyces*, *Saksenaea*, *Mucor*, and *Absidia* also are occasionally responsible for this infection. The organism in tissue is composed of broad, rarely septate hyphae of uneven diameter (6 to 50 um). The organisms are inexplicably difficult to grow from infected tissue. When growth does take place, it is rapid and profuse on most media at room temperature. Identification is based on the gross and microscopic appearance of the mold.

Zygomycosis is a term that includes mucormycosis and entomophthoramycosis. The latter is a tropical infection of the subcutaneous tissue or paranasal sinuses caused by species of *Basidiobolus* and *Conidiobolus*, respectively.

Epidemiology and Pathology

Rhizopus and *Rhizomucor* species are ubiquitous, appearing on decaying vegetation, dung, and foods of high sugar content. Mucormycosis is uncommon and is largely confined to patients with serious preexisting diseases. Mucormycosis originating in the paranasal sinuses and nose predominantly affects patients with poorly controlled diabetes mellitus. Patients who have undergone organ transplantation, who have a hematologic malignancy, or who are receiving long-term deferoxamine therapy are predisposed to mucormycosis of either sinus or lung. Gastrointestinal mucormycosis occurs in a variety of conditions, including uremia, severe malnutrition, and diarrheal diseases. The infection is acquired from nature, with no person-to-person spread. In all forms of mucormycosis, vascular invasion by hyphae is a prominent feature. Ischemic or hemorrhagic necrosis is the foremost histologic finding.

Clinical Manifestations

Mucormycosis originating in the nose and paranasal sinuses produces a characteristic clinical picture. Low-grade fever, dull sinus pain, and sometimes nasal congestion or a thin, bloody nasal discharge are followed in a few days by double vision, increasing fever, and obtundation. Examination reveals a unilateral generalized reduction of ocular motion, chemosis, and proptosis. The nasal turbinates on the involved side may be dusky red or necrotic. A sharply delineated area of necrosis, strictly respecting the midline, may appear in the hard palate. The skin of the cheek may become inflamed. Fungal invasion of the globe or ophthalmic artery leads to blindness. Opacification of one or more sinuses is detected by computed tomography (CT) or by magnetic resonance imaging (MRI). Carotid arteriography may show invasion or obstruction of the carotid siphon. Coma is due to direct invasion of the frontal lobe. Early symptoms mimic those of bacterial sinusitis. Clouding of the sensorium may be attributed to diabetic acidosis. Cavernous sinus thrombosis may be considered when orbital invasion occurs. Without treatment, the patient may die after an interval ranging from a few days to a few weeks.

Pulmonary mucormycosis manifests as progressive severe pneumonia accompanied by high fever and toxicity. The necrotic centre of large infiltrates may cavitate. Hematogenous spread to other areas of the lung, as well as to the brain and other organs, is common. Survival beyond 2 weeks is unusual. Gastrointestinal invasion presents as one or more ulcers that tend to perforate. Hematogenous dissemination can originate from the gastrointestinal tract, lung, or paranasal sinuses. Sometimes no portal of entry can be found.

Diagnosis

CT or MRI is very helpful in assessing the extent of sinusitis before surgery and in evaluating the patient afterward. CT is better for detecting bony erosion; MRI better visualizes extension into the frontal lobe or carotid artery in the siphon. Lesions of the lung and craniofacial structures are best diagnosed by biopsy and histologic section. Cultural confirmation should be attempted. Wet smear of crushed tissue can provide a rapid diagnosis. Cultures of blood and cerebrospinal fluid are negative. Smear and culture of sputum may be positive during cavitation of a lung lesion.

Treatment

Regulation of diabetes mellitus and a decrease in the dose of immunosuppressive drugs facilitate the treatment of mucormycosis. Extensive debridement of craniofacial lesions appears to be very important. Orbital exenteration may be required. Intravenous amphotericin B is clearly of value in craniofacial mucormycosis and should be employed in the other forms of mucormycosis as well. The maximal tolerated doses are given until progression is halted. With the deoxycholate formulation, 1 to 1.5 mg/kg daily is indicated. Therapy is continued for a total of 10 to 12 weeks. Azoles are of no value. Appropriate management results in cure of about half of craniofacial infections. The survival of patients with pulmonary, gastrointestinal, or disseminated mucormycosis is rare.

MISCELLANEOUS MYCOSES AND ALGAL INFECTIONS

Chromoblastomycosis

This chronic subcutaneous mycosis, rarely seen in the United States, presents as a verrucoid, ulcerated, or crusted skin lesion. The disease follows the introduction of any of several fungi into subcutaneous tissue by thorns or bits of vegetation. The infection spreads over ensuing months and years to contiguous tissue, causing few symptoms. The appearance of thick-walled, dark-coloured, rounded forms ("copper pennies") in histopathologic section is diagnostic. Surgical excision is the treatment of choice. Itraconazole has ameliorated some relatively small and incompletely excised lesions.

Dermatophytosis

Dermatophytosis, also known as ringworm or tinea, is a chronic fungal infection of the skin, hair, or nails.

Etiology Species of *Trichophyton, Microsporum,* and *Epidermophyton* are called *dermatophytes.* These organisms grow in and remain confined to the keratinous structures of the body. Other mycoses, such as candidiasis, pityriasis versicolour, and tinea nigra, sometimes include fungal invasion of keratinous structures but traditionally are not called dermatophytoses.

Pathology and Pathogenesis: Dermatophyte species are referred to as *anthropophilic*, *zoophilic*, or *geophilic*, depending on whether their usual reservoir in nature appears to be humans, animals, or soil, respectively. The infectivity of organisms from all these sources is low, and outbreaks are largely confined to occasional clusters of cases of scalp infection in children. Acquisition of a dermatophytosis appears to be favoured by minor trauma, maceration, and poor hygiene of the skin. Infection does not seem to confer solid immunity: Repeated infection with the same species is common, particularly with anthropophilic species. The infrequency of scalp infection among adults has been attributed to local factors rather than immunity.

Invasion of the stratum corneum by dermatophytes may cause inflammation that is either mild or (particularly with zoophilic fungi) intense. Shedding of the stratum corneum is increased by inflammation. To the extent that fungal growth cannot keep up with shedding, inflammation may help terminate infection. Conversely, infection is probably favoured when shedding is reduced by treatment with glucocorticoids and cytotoxic drugs. Antifungal drugs interfere with the ability of fungal growth to keep up with shedding.

Clinical Manifestations: The disease varies with the site of infection and the fungal species involved. Foot infection (athlete's foot, tinea pedis) may present as fissuring of the toe webs, scaling of the plantar surfaces, or vesicles around the toe webs and soles. Interdigital lesions may be pruritic or, when bacterial superinfection occurs, may be painful. Hand infection is less common but resembles foot infection.

Scalp dermatophytosis (tinea capitis) is characterized by areas of alopecia and scaling. In so-called endothrix infection, the hair shaft breaks off at the skin surface, leaving the hairs

visible as black dots in the scalp. Some forms of scalp infection include an area of intense boggy suppuration called a *kerion*.

Dermatophytosis of the glabrous skin (tinea corporis) presents as circumscribed lesions with a wide variety of appearances, including scales, vesicles, and pustules. Inflammation may be minimal or intense. Central healing of less inflamed lesions may take place. The serpiginous border of inflammation is the source of the name *ringworm*.

Dermatophytosis of the bearded area (tinea barbae) appears as a pustular folliculitis. Onychomycosis (tinea unguium) presents as a white discolouration of the nails or as thickening, chalkiness, and crumbling of the nails. Peeling and fissuring of paronychial nail folds or keratotic debris under the nail edge also may be evident.

Diagnosis: Discoloured hairs, scales, and keratotic debris under infected nails should be collected for KOH smear and culture. In the scraping of skin lesions, a drop of water on the skin site may keep the removed scales from flying off and thus may aid in their collection. Culture is important in distinguishing dermatophytes from *Candida* and fungal saprophytes growing in keratinaceous debris.

Treatment

Noninflammatory lesions of the trunk, groin, hands, and feet usually respond to twice-daily applications of clotrimazole, miconazole, ketoconazole, econazole, naftifine, terbinafine, or ciclopirox olamine cream. Hyperkeratotic lesions of the palms and soles respond slowly to these agents and may benefit from Whitfield's ointment initially to thin the keratin. Ointment should not be used between the toes, in the groin, or in the gluteal crease because maceration promotes bacterial infection.

Ringworm that is moderately severe, that is unresponsive to topical therapy, or that involves the scalp, nails, or bearded area should be treated systemically. Once-daily therapy with itraconazole (200 mg), terbinafine (250 mg), microcrystalline griseofulvin (500 mg), or ultramicrocrystalline griseofulvin

(375 mg) is effective. Treatment must be continued until all infected keratin is gone. Cutting off infected hair and cleansing interdigital webs can expedite cure. Secondary bacterial infection of the foot may require soaks or antibacterial agents. The likelihood of relapse of dermatophyte foot infections may be decreased by keeping the feet clean and dry. For nail infections, itraconazole or terbinafine is preferred. In distal subungual onychomycosis, a single course of either drug results in initial improvement in half of patients, of whom half relapse. Results are better with fingernails than with toenails and for more distal rather than lateral nail involvement. To save money, itraconazole can be given as a double dose (400 mg) for 1 week each month with only marginal loss of efficacy. The duration of therapy is 2 to 3 months for fingernails and 4 to 6 months for toenails.

Protothecosis

Prototheca species are ubiquitous achlorophyllic algae that enter the skin through contaminated wounds and cause localized infections in the olecranon bursa, skin, subcutaneous tissue, tendon sheaths, or deeper tissue. Diagnosis is based on culture or histopathologic demonstration of sporangia with endospores in tissue. Surgical debridement and treatment with intravenous amphotericin B are useful.

Fusariosis

Fusarium species can cause localized or hematogenously disseminated infection. Localized infection results from contaminated wounds. Almost all patients with hematogenously disseminated infection are severely immunosuppressed and profoundly neutropenic. Skin lesions occur in two-thirds of patients. Several painful red indurated lesions appear on the extremities or sometimes the trunk. These lesions often develop an ecchymotic centre that ulcerates. A portal of infection is not usually apparent. Blood cultures have been positive in 59 per cent of cases. Amphotericin B is probably the drug of choice for the treatment of fusariosis, but recovery depends on the diminution of neutropenia.

Malassezia Infection (Pityriasis)

Malassezia furfur is part of the normal flora of the human skin but can cause tinea (pityriasis) versicolour or catheter-acquired sepsis. Tinea versicolour appears as asymptomatic, well-delineated, hyperpigmented or hypopigmented macules centreed on the upper trunk and upper arms. Confluent lesions may cover large areas, making the border difficult to find. A fine "branny" scale or folliculitis is sometimes visible. When examined microscopically by KOH mount, skin sections are seen to contain characteristic round and elongated cells. On inspection with Wood's light, lesions either do not fluoresce or appear yellow-green. *Erythrasma* resembles tinea versicolour but is characterized by gram-positive bacilli on smear and coral-red fluorescence. Azole creams are effective for the treatment of small areas of tinea versicolour; however, the application of selenium sulfide shampoo (Selsun) for 10 min daily, followed by showering to remove the shampoo, is more practical for large areas. Itraconazole is also effective. Catheter-acquired sepsis due to *M. furfur* develops in patients (particularly neonates) receiving intravenous lipid. The organism requires special culture conditions for growth, and the infection is cured by catheter removal.

MYCETOMA

Etiology: *Actinomycetoma* refers to infection by actinomycetes of the genera *Nocardia*, *Nocardiopsis*, *Streptomyces*, and *Actinomadura*. *Eumycetoma* is caused by true fungi of many different genera. The predominant agent varies with the locality.

Pathogenesis and Pathology: The pathogens live in the soil and enter the skin through minor trauma. The most common site of infection is the foot. The infection runs a relentless course over many years, with destruction of contiguous bone and fascia. Grains are found in purulent foci surrounded by fibrosis and a mononuclear cell inflammatory response.

Clinical Manifestations: Mycetoma is a chronic suppurative infection originating in subcutaneous tissue and characterized by the presence of grains, which are tightly clumped colonies of the causative agent. The infected site is characterized by painless swelling, woody induration, and sinus tracts that discharge pus intermittently. Systemic symptoms do not develop, and spread to distant sites in the body does not take place.

Diagnosis: Although the clinical picture is characteristic, mycetoma is sometimes confused with chronic osteomyelitis or botryomycosis. The diagnosis requires demonstration of grains in pus from the draining sinus or in biopsy sections. Many histologic sections may need to be examined to locate a grain.

Treatment

Actinomycetoma may respond to prolonged combination chemotherapy with streptomycin and either dapsone or trimethoprim-sulfamethoxazole. Eumycetoma rarely responds to chemotherapy; some cases caused by *Madurella mycetomatis* have appeared to respond to ketoconazole or itraconazole.

Paracoccidioidomycosis

Etiology: Formerly called *South American blastomycosis*, this mycosis is caused by *Paracoccidioides brasiliensis*. A dimorphic fungus, *P. brasiliensis* grows as a budding yeast in tissue but may be grown as either a yeast or a mold on culture medium. The organism is identified by its gross and microscopic appearance.

Pathogenesis and Pathology: Infection is thought to be acquired by inhalation of spores from environmental sources, possibly soil. Pulmonary infection produces few symptoms initially. Hematogenous spread to the mucous membranes of the mouth and nose, the lymph nodes, and other sites causes patients to seek medical attention. In fatal cases, the infection spreads to the adrenals, the gastrointestinal tract, and many other viscera.

Clinical Manifestations: Common signs include indurated ulcers of the mouth, oropharynx, larynx, and nose; enlarged and draining lymph nodes; lesions of the skin and genitalia; and productive cough, weight loss, dyspnea, and sometimes fever. Paracoccidioidomycosis is acquired only in South America, Central America, and Mexico, but its extreme indolence may delay its recognition until many years after the patient has left the endemic area. Chest radiography most often shows bilateral patchy pneumonia.

Diagnosis: Cultures of sputum, pus, and mucosal lesions are often diagnostic. The diagnosis can be made by smear or histologic section, although confirmation by culture is preferable. Serologic tests are useful in suggesting the diagnosis and monitoring the response to therapy.

Treatment

Relatively mild cases of paracoccidioidomycosis may be cured by 1 year of treatment with oral ketoconazole or itraconazole (200 to 400 mg daily). More advanced cases are treated with intravenous amphotericin B followed by itraconazole.

Phaeohyphomycosis

This is the name given to infections caused by fungi with dark-walled hyphae, excluding those given conventional names like chromoblastomycosis. Although an extraordinary variety of fungi and clinical syndromes are encompassed by this definition, most patients have brain abscess, subcutaneous abscess, or allergic fungal sinusitis. Most of the brain abscesses are due to *Cladophialophora bantiana*, *Ochroconis gallopavum*, *Exophiala dermatitidis*, *Bipolaris* species, and *Ramichloridium mackenziei*. Patients are previously healthy. Subcutaneous abscesses are usually single, arise at the site of minor trauma, and occur in both immunosuppressed and immunocompetent individuals. A large number of dematiaceous (dark-walled) mold species cause subcutaneous phaeohyphomycosis as well as allergic fungal sinusitis. The latter entity develops in patients with allergic rhinitis and presents as an expanding mucoid mass in one or more paranasal sinuses. The tenacious mucus contains eosinophils, Charcot-Leyden crystals, and occasional hyphae. Surgical excision of phaeohyphomycotic lesions is important. Antifungal therapy may retard recurrences but is of little value unless surgical excision has been performed.

Pseudallescheriasis

Etiology: Also called *Petriellidium boydii, Pseudallescheria boydii* is a mold frequently found in soil. When the fungus is isolated in the imperfect state, it is called *Scedosporium apiospermum*.

Pathogenesis and Pathology: Wind-borne spores of *P. boydii*, arising from the soil, are the presumed source of infection. The fungus grows as a mold within tissue, causing necrosis and abscess formation.

Clinical Manifestations: *P. boydii* resembles *Aspergillus* in its ability to colonize the endobronchial tree, to form fungus balls in the lungs or paranasal sinuses, and to invade the cornea or globe of the eye, the soft tissues, the joints, or the bones after trauma or surgery and in its propensity to invade the lungs and paranasal sinuses of immunosuppressed hosts, including patients with AIDS. Hyphae of *P. boydii* in tissue may be difficult to distinguish from those of *Aspergillus*. Infection with *P. boydii* is much less common than that with *Aspergillus*. Intravascular hyphae, a hallmark of invasive aspergillosis, are also found in pseudallescheriasis. Near-drowning in polluted water has led to severe *P. boydii* pneumonia, often with dissemination and fatal brain abscesses.

Diagnosis: Demonstration of hyphae in tissue and culture confirmation are required for diagnosis.

Treatment

Itraconazole at the maximal tolerated doses is the regimen of choice. Surgical drainage or debridement can be helpful. The prognosis is poor.

Scedosporium prolificans, a fungus closely related to *P. boydii*, has caused infections in bones, joints, or soft tissue, usually after trauma. These infections have responded to surgical debridement. Disseminated infection with *S. prolificans* in immunosuppressed patients has been fatal. The response to treatment with all antifungal agents has been poor.

Sporotrichosis

Etiology: *Sporothrix schenckii* lives as a saprophyte on plants in many areas of the world. In nature and on culture at room temperature, the fungus grows as a mold; within host tissue or at 37°C on enriched media, it grows as budding yeast. It is identified by its appearance in mold and yeast forms.

Pathogenesis and Pathology: Infection results from the inoculation of *S. schenckii* into subcutaneous tissue through minor trauma. Nursery workers, florists, and gardeners acquire the illness from roses, sphagnum moss, and other plants. Infection may be limited to the site of inoculation (plaque sporotrichosis) or extend along proximal lymphatic channels (lymphangitic sporotrichosis). Spread beyond an extremity the usual site of infection is rare, and hematogenous dissemination from the skin remains unproven. The portal for osteoarticular, pulmonary, and other extracutaneous forms of sporotrichosis is unknown but is likely the lung.

Untreated sporotrichosis persists for months. The inflammatory response includes both the clustering of neutrophils and a marked granulomatous response with epithelioid cells and giant cells.

Clinical Manifestations: In lymphangitic sporotrichosis, which is by far the most common manifestation, a nearly painless red papule forms at the site of inoculation. Over the next several weeks, similar nodules form along proximal lymphatic channels. The nodules intermittently discharge small amounts of pus. Ulceration may occur. The proximal extension of these lesions, often with skip areas, is quite distinctive but may be mimicked by lesions of *Nocardia brasiliensis*, *Mycobacterium marinum*, or (in rare cases) *Leishmania brasiliensis* or *Mycobacterium kansasii*.

Plaque sporotrichosis manifests as a nontender red maculopapular granuloma confined to the site of inoculation.

Osteoarticular sporotrichosis presents as mono- or polyarticular arthritis of indolent onset and progression over months or years, involving the elbows, knees, wrists, ankles, and (rarely) smaller joints of the extremities. Periarticular bone develops areas of demineralization detectable on x-ray, and draining sinuses may appear over joints and bursae. Hematogenous spread to the skin may take place during polyarticular disease, but none of the skin lesions shows lymphangitic spread. Immunosuppression, including that due to advanced infection with HIV, predisposes to hematogenous spread. Pulmonary sporotrichosis usually presents as a single chronic cavitary upper-lobe lung lesion. Chronic meningitis can develop in the absence of skin or lung lesions. *S. schenckii* is difficult to recover from cerebrospinal fluid.

Diagnosis: Culture of pus, joint fluid, sputum, or a skin biopsy specimen is the preferred method of diagnosis. The appearance of *S. schenckii* in tissue is quite variable. In skin lesions, the organisms are hard to find.

Treatment

Cutaneous sporotrichosis can be cured with a saturated solution of potassium iodide given orally in increasing divided doses of up to 4.5 to 9 mL/d for adults, as tolerated. Gastrointestinal disturbance or acneiform rash over the cape area and face is common, but therapy should be continued for 1 month after the resolution of all lesions. Itraconazole (100 to 200 mg daily) is an effective and better-tolerated alternative. Extracutaneous sporotrichosis rarely responds to iodides, but more than half of cases have been cured by prolonged courses of intravenous amphotericin B. Itraconazole (200 mg once or twice daily) is effective in some cases of extracutaneous sporotrichosis.

Trichosporonosis

A recent change in taxonomy of the genus *Trichosporon* has moved most of the agents causing deep infections from *T. beigelii* into the species *T. asahii*, with a few categorized as *T. mucoides*. White piedra of the scalp is caused by *T. ovoides* and that of the pubic hair by *T. inkin. T cutaneum* and *T. asteroides* cause superficial infections. Most of what is currently known about *Trichosporon* infections is not species specific, so the following description refers to *T. beigelii. T. capitatum*, which causes disseminated infection in patients with neutropenia, was previously reclassified as *Blastoschizomyces capitatus* and will not be covered here.

T. beigelii can colonize the human gastrointestinal tract and skin and can enter the bloodstream of patients with severe neutropenia through an inapparent source. Hematogenously disseminated infection is manifested by fever and often by the development of several erythematous or purpuric tender papules anywhere on the body. Lesions can form large, tense hemorrhagic bullae. In some patients, native or prosthetic cardiac valves become infected. In tissue, hyphae and yeastlike cells are seen. Amphotericin B is probably the drug of choice for treatment, but recovery is dependent on the return of bone marrow function.

PNEUMOCYSTIS CARINII INFECTION

Definition and Description

Pneumocystis carinii is an opportunistic pathogen whose natural habitat is the lung. The organism is an important cause of pneumonia in the compromised host.

Although the taxonomic status of *P. carinii* has long been controversial, molecular studies during the past decade have clearly placed the organism among the fungi. This classification is based on analysis of gene sequences for ribosomal RNA, mitochondrial proteins, and major enzymes. The cell wall of *P. carinii* contains b-1,3-glucan; drugs that inhibit b-glucan synthesis in fungi are highly active against *P. carinii* in animal models. However, in contrast to most fungi, *P. carinii* lacks ergosterol and is not susceptible to antifungal drugs that inhibit ergosterol synthesis.

Study of the basic biology of *P. carinii* has been severely hampered by the lack of a reliable in vitro cultivation system. Major developmental stages of the organism include the small

(1- to 4-um) pleomorphic trophozoite or trophic form; the 5- to 8-um cyst, which has a thick cell wall and contains up to eight intracystic bodies; and the precyst, an intermediate stage. The life cycle of *P. carinii* probably involves asexual replication by the trophic form and sexual reproduction by the cyst, which ends in release of the intracystic bodies; an intracellular stage has not been identified. Ultrastructurally, *P. carinii* has a primitive organelle system, but little is known about its metabolism.

P. carinii contains two prominent antigen groups. The 95- to 140-kDa major surface glycoprotein (MSG) complex represents a family of proteins encoded by multiple genes. The MSG complex is highly immunogenic, contains shared and species-specific antigenic determinants, and exhibits protective B and T cell epitopes in animal models. The MSG complex plays a pivotal role in the host-parasite relationship in *P. carinii* infection. It facilitates adherence to host proteins via extracellular matrix proteins, surfactant proteins, and the mannose receptor; and its ability to undergo antigenic variation may represent a mechanism by which *P. carinii* evades the host immune response. The other antigen, which migrates as a band of 35 to 55 kDa, is the most common antigen recognized by the host and thus may serve as a marker of infection.

Epidemiology

P. carinii has a worldwide distribution among humans and has been detected in a variety of animals. The organisms found in these hosts are morphologically identical, but recent studies have revealed a high degree of genetic diversity and host specificity. Serologic surveys indicate that most healthy children have been exposed to the organism by 3 to 4 years of age. Animal model experiments have demonstrated that *P. carinii* is transmitted by the airborne route. Person-to-person transmission has been suggested by the occurrence of outbreaks of pneumocystosis among institutionalized debilitated infants and in hospitals caring for immunosuppressed patients. On the basis of animal studies, the incubation period is thought to be 4 to 8 weeks.

Pathogenesis and Pathology

The host factors that predispose to the development of pneumocystosis involve defects in cellular and humoural immunity. People at risk for the disease include patients infected with HIV; persons receiving immunosuppressive therapy (particularly glucocorticoids) for cancer, organ transplantation, and other disorders; children with primary immunodeficiency diseases; and premature malnourished infants. The central role of CD4+ cells in host resistance to *P. carinii* has been shown by research in experimental animals and by studies that have correlated the risk of pneumocystosis in HIV-infected patients with CD4+ cell counts. Evidence supporting the importance of impaired humoural immunity consists of the occurrence of pneumocytosis in patients and animals with B cell defects and the beneficial effect of passively administered antibodies.

The principal host effector cells against *P. carinii* are alveolar macrophages, which ingest and kill the organism, releasing a variety of inflammatory mediators. Tumor necrosis factor a and interleukin (IL) 1 are important in the early host defences against *P. carinii*, but the role of other cytokines is less clear. Recent evidence suggests that HIV alters the mannose receptor-mediated binding and phagocytosis of *P. carinii*.

After being inhaled, *P. carinii* takes up residence in the alveoli, where it attaches tightly to type I cells but maintains an extracellular existence. In some cases, the organism remains in the host for long periods, and pneumonia develops by reactivation of latent infection; in other cases, pneumonia arises from a new bout of infection. As the immune system of the host becomes compromised, *P. carinii* organisms propagate and gradually fill the alveoli. This scenario is accompanied by a complex series of events that result in increased alveolar-capillary permeability and damage to type I cells. Surfactant abnormalities include a fall in bronchoalveolar lavage (BAL) fluid phospholipids and an increase in surfactant proteins A and D. Contributions of the host inflammatory response to lung injury are suggested by the correlation of increased IL-8 levels and neutrophil counts in BAL fluid from patients with relatively severe disease.

On lung sections stained with hematoxylin and eosin, the alveoli are filled with a typical foamy, vacuolated exudate. Severe disease may include interstitial edema, fibrosis, and hyaline membrane formation. The host inflammatory changes usually consist of hypertrophy of alveolar type II cells, a typical reparative response, and a mild mononuclear cell interstitial infiltrate. Malnourished infants display an intense plasma cell infiltrate that gave the disease its early name: interstitial plasma cell pneumonia.

Clinical Features

Patients with *P. carinii* pneumonia develop dyspnea, fever, and nonproductive cough. Symptoms in non-HIV-infected patients often begin after the glucocorticoid dose has been tapered and typically last 1 to 2 weeks. HIV-infected patients are usually ill for several weeks or longer and have relatively subtle manifestations. However, the clinical picture in individual patients is quite variable, and a high index of suspicion and elicitation of a careful history are key factors in early detection.

Physical findings include tachypnea, tachycardia, and cyanosis, but lung auscultation reveals few abnormalities. The white blood cell count is variable and is usually governed by the patient's underlying disease. Assessment of arterial blood gases demonstrates hypoxia, an increased alveolar-arterial oxygen gradient, and respiratory alkalosis. There also may be changes in pulmonary function test values (diffusing capacity) and increased uptake with nuclear imaging techniques (gallium scan). Elevated serum concentrations of lactate dehydrogenase (LDH) have been reported; they probably reflect lung parenchymal damage but are not specific to *P. carinii* infection. In general, laboratory abnormalities are less severe in HIV-infected patients than in non-HIV-infected patients.

The classic findings on chest radiography consist of bilateral diffuse infiltrates beginning in the perihilar regions, but various atypical manifestations (nodular densities, cavitary lesions) have also been reported. Patients who receive aerosolized pentamidine have an increased frequency of upper-lobe infiltrates and pneumothorax. Early in the course of pneumocystosis, the chest radiograph may be normal.

Although *P. carinii* usually remains confined to the lungs, cases of disseminated infection have occurred in both HIV-infected and non-HIV-infected patients. One risk factor for extrapulmonary spread in patients with HIV is the administration of aerosolized pentamidine. The most common sites of extrapulmonary involvement are the lymph nodes, spleen, liver, and bone marrow. Clinical manifestations range from incidental findings at autopsy to specific organ involvement. Histopathologic examination reveals the presence of *P. carinii* and the characteristic associated foamy material. Treatment for the extrapulmonary forms of pneumocystosis is the same as that for pneumonia.

Diagnosis

Because the clinical picture of *P. carinii* infection can be produced by many other infectious and noninfectious agents, the diagnosis must be based on specific identification of the organism. A definitive diagnosis is made by histopathologic staining. Traditional stains have included reagents such as methenamine silver, toluidine blue, and cresyl echt violet, which selectively stain the wall of *P. carinii* cysts, and reagents such as Wright-Giemsa, which stain the nuclei of all developmental stages. Other reagents include nonspecific fluorochrome stains (calcofluor white) and Papanicolaou's stain. Immunofluorescence with monoclonal antibodies is more sensitive than histologic staining but is also more expensive. DNA amplification by the polymerase chain reaction offers the greatest sensitivity and may find a place in the routine diagnosis of *P. carinii* when commercial kits become available.

The successful diagnosis of pneumocystosis depends upon the collection of proper specimens. In general, the yield from different diagnostic procedures is higher in HIV-infected patients than in non-HIV-infected patients because of the higher organism burden in the former group. Sputum induction has gained popularity as a simple, noninvasive technique; this procedure requires trained and dedicated personnel, and its success has varied at different institutions. Fibreoptic bronchoscopy with BAL, which is more sensitive than sputum induction, remains the mainstay of *P. carinii* diagnosis. This procedure also provides information about the organism burden, the host inflammatory response, and the presence of other opportunistic infections. Transbronchial biopsy and open lung biopsy, which are the most invasive procedures, are reserved for situations in which a diagnosis cannot be made by BAL.

Course And Prognosis

In the typical case of untreated *P. carinii* pneumonia, progressive respiratory embarrassment leads to death. Therapy is most effective when instituted early in the course of the disease, before there is extensive alveolar damage. If induced sputum is nondiagnostic and BAL cannot be performed in a timely manner, it is reasonable to begin empiric therapy with drugs active against *P. carinii*. However, this practice does not obviate the need for a specific etiologic diagnosis. With improvements in management, the case-fatality rate has been lowered to about 15 per cent in HIV-infected patients but remains high (40 per cent) in non-HIV-infected patients. The most widely used prognostic indicators have been the arterial oxygen pressure and the alveolar-arterial oxygen gradient. Other factors that may influence survival include neutrophil counts and IL-8 levels in BAL fluid, chest radiographic abnormalities, serum LDH and albumin levels, and the expertise of the hospital in caring for patients with HIV infection. Concurrent pulmonary infections complicate management, but the presence of cytomegalovirus usually does not affect the outcome of pneumocystosis.

Treatment

Trimethoprim-sulfamethoxazole (TMP-SMZ), which acts by inhibiting folic acid synthesis, is considered the drug of choice for all forms of pneumocystosis. The daily dosage, administered orally or intravenously in three or four divided doses, is 15 to 20 mg TMP/kg and 75 to 100 mg SMZ/kg. Therapy is continued for 14 days in non-HIV-infected patients and for 21 days in persons infected with HIV. Since HIV-infected patients respond more slowly than non-HIV-infected patients, it is prudent to wait at least 7 days after the initiation of treatment before concluding that therapy has failed. The addition of drugs to an existing regimen is no more effective than switching regimens and may increase the risk of toxicity. TMP-SMZ is well tolerated by non-HIV-infected patients, but more than half of HIV-infected patients experience serious adverse reactions, including fever, rash, neutropenia, thrombocytopenia, hepatitis, and hyperkalemia.

Several alternative regimens are available for the treatment of mild to moderate cases of *P. carinii* pneumonia: TMP (15 mg/kg per day orally) plus dapsone (100 mg/d orally), clindamycin (600 mg every 6 h intravenously or 300 to 450 mg every 6 h orally) plus primaquine (15 to 30 mg of base per day orally), or atovaquone alone (750 mg twice daily orally). Dapsone and primaquine should be used with caution in patients with glucose-6-phosphate dehydrogenase deficiency.

Two alternative drugs are available for the treatment of moderate to severe forms of pneumocystosis. Pentamidine, which has been used against *P. carinii* for many years, is administered as a single daily dose of 4 mg/kg by slow intravenous infusion. Pentamidine is highly toxic; its major side effects are hypotension, cardiac arrhythmias, dysglycemias, azotemia, electrolyte changes, and neutropenia. Trimetrexate is administered intravenously as a single daily dose of 45 mg/m2; in conjunction with trimetrexate therapy, folinic acid is given orally or intravenously at a dose of 20 mg/m2 every 6 h to prevent bone marrow suppression.

Patients with HIV frequently experience deterioration in respiratory function shortly after receiving anti-*P. carinii* drugs. Several studies have shown that the administration of glucocorticoids to patients with HIV and moderate to severe pneumocystosis can prevent this problem and improve the rate of survival. The administration of steroids should be started early in the course of the illness (usually when antimicrobial drugs are begun) for maximal benefit; the recommended regimen is 40 mg of prednisone orally twice daily on days 1 to 5, 40 mg/d on days 6 to 10, and 20 mg/d on days 11 to 20. This regimen has generally proven to be safe despite concern about its effects on other opportunistic infections. The use of steroids as adjunctive therapy in HIV-infected patients with mild pneumocystosis or in non-HIV-infected patients remains to be evaluated.

Prevention

Primary prophylaxis is indicated for HIV-infected patients at high risk of developing pneumocystosisthat is, those who have CD4+ cell counts of < 200/uL, unexplained fever [> 37.8°C (100° F)] for ³ 2 weeks, or a history of oropharyngeal candidiasis. Guidelines for the administration of primary prophylaxis to other immunocompromised hosts are less clear. Secondary prophylaxis is indicated for all patients who have recovered from *P. carinii* pneumonia. Among HIV-infected patients, the risk of recurrent episodes of pneumocystosis is high and lifelong; among non-HIV-infected patients, the risk is lower and exists for as long as the immunosuppressive condition persists.

Several antimicrobial drugs are effective in preventing pneumocystosis, although some concern has been raised about possible resistance. One double-strength tablet of TMP-SMZ (160 mg TMP, 800 mg SMZ) per day is the prophylactic regimen of choice. The major limitation of TMP-SMZ treatment is the high frequency of adverse reactions among HIV-infected patients. Recommended alternative regimens include TMP-SMZ at a reduced dose (80 mg TMP, 400 mg SMZ) or frequency (3 times per week), dapsone alone at a daily oral dose of 100 mg, dapsone at a dose of 50 mg/d combined with weekly oral doses of pyrimethamine (50 mg) and folinic acid (25 mg), pentamidine at a monthly dose of 300 mg administered by Respigard nebulizer, and atovaquone at an oral dose of 1500 mg/d.

Although there are no specific recommendations for preventing the spread of *P. carinii* in health care facilities, it seems prudent to prevent direct contact between patients with pneumocystosis and other susceptible hosts.

